Contributions to Stochastics

Contributions to Stochastics

EDITED BY
N. VENUGOPAL
FSS (Lond.)

PUBLISHING FOR ONE WORLD

NEW AGE INTERNATIONAL (P) LIMITED, PUBLISHERS
New Delhi • Bangalore • Chennai • Cochin • Guwahati • Hyderabad
Jalandhar • Kolkata • Lucknow • Mumbai • Ranchi
Visit us at **www.newagepublishers.com**

Published by New Age International (P) Ltd., Publishers
First Edition: 1992
Reprint: 2009

Branches:

- 36, Malikarjuna Temple Street, Opp. ICWA, Basavanagudi, **Bangalore**. ✆ (080) 26677815
- 26, Damodaran Street, T. Nagar, **Chennai**. ✆ (044) 24353401
- CE-39/1016, Carrier Station Road, Near Paulson Park Hotel, Ernakulam South, **Cochin.** ✆ (0484) 4051303
- Hemsen Complex, Mohd. Shah Road, Paltan Bazar, Near Starline Hotel, **Guwahati**. ✆ (0361) 2543669
- No. 105, 1st Floor, Madhiray Kaveri Tower, 3-2-19, Azam Jahi Road, Nimboliadda, **Hyderabad**. ✆ (040) 24652456
- RDB Chambers (Formerly Lotus Cinema) 106A, 1st Floor, S.N. Banerjee Road, **Kolkata**. ✆ (033) 22275247
- 16-A, Jopling Road, **Lucknow**. ✆ (0522) 2209578
- 142C, Victor House, Ground Floor, N.M. Joshi Marg, Lower Parel, **Mumbai**. ✆ (022) 24927869
- 22, Golden House, Daryaganj, **New Delhi**. ✆ (011) 23262370, 23262368

ISBN: 978-81-224-0417-3

Rs. 495.00

C-09-07-3870

Printed in India at Glorious Printers, Delhi.

PUBLISHING FOR ONE WORLD
NEW AGE INTERNATIONAL (P) LIMITED, PUBLISHERS
4835/24, Ansari Road, Daryaganj, New Delhi-110002
Visit us at **www.newagepublishers.com**

Foreword

Professor Kandula Nagabhushanam
(1908–1979)

Professor K. Nagabhushanam was one of the two inspiring teachers I had when I was pursuing graduate studies in mathematics at the Andhra University, Waltair. He and Professor V. Ramaswamy not only taught us mathematics but also prepared us to think in terms of mathematics and to use mathematics as an abstract logical method for solving complex problems in any field of inquiry. This training was a great asset to me when I started on my research career. I had kept in touch with Professor Nagabhushanam after I left the Andhra University, as he was keenly interested in my activities and often encouraged me in my research work. It is, indeed, a great honor to contribute to the memorial volume dedicated to my respected teacher.

C. R. Rao

LIFE AND WORKS OF PROFESSOR KANDULA NAGABHUSHANAM (1908–1979)

A BIOGRAPHICAL SKETCH

N. VENUGOPAL

AND

K. D. RAMA RAO

Prof. K. Nagabhushanam was born on February 1, 1908 at Tenali in Andhra Pradesh (India). He had his early education at several schools in Guntur District (Andhra Pradesh) before he joined P.R. College, Kakinada for his B.A. Degree. He obtained the degree in 1928 securing the First place and was awarded Three Gold Medals by the Andhra University (Waltair). For higher studies, he went to Madras and joined the Presidency College. He obtained his M.A. Degree in Mathematics (1931), and stood second in the University. While studying at the Madras University, he also obtained Diplomas in French and German. From 1926 to 1934, he was recipient of three scholarships and fellowships of the Madras and Andhra Universities.

After a brilliant academic career, he joined Andhra University, in 1934, as Lecturer in the Department of Mathematics. He served that department for 12 years and thereafter for one year worked in the Mathematical Physics Department. In 1946, he was promoted and appointed Reader in the Department of Mathematics. After he had served as Reader for four years, he was deputed by the Andhra University to the Institute of Mathematical Statistics, University of Stockholm, Sweden, to pursue higher studies in the field of Statistics. He worked under the supervision of Prof. Harald Cramer, the internationally renowned teacher and researcher in Statistics, and obtained Fil. Lic. (1950) and D. Phil. (1951) Degrees in Statistics from the University of Stockholm. After returning to the Andhra University, in 1952, he started the Department of Statistics of the Andhra University and thus became a pioneer in this new and important discipline.

Dr. K. Nagabhushanam was Professor and the Head of the Department of Statistics from 1952 to 1965 and again from 1966 to 1967. During 1965–66, he went abroad, as a Visiting Professor of Mathematics, to the University of Alberta, Edmonton, Canada. Ill-health forced him to cut short not only his foreign assignment but also to give up the administrative responsibilities as the Head of Department. Even so, he devoted his time entirely to teaching and research, till he retired from service in 1968. Even after his retirement, he served the University and the Department as Professor Emeritus and Professor under the University Grants Commission's Retired Teachers' Scheme for another three years.

Prof. Nagabhushanam's works on Non-stationary Processes, particularly related to spectrum, had won him international reputation. Some of his results are well cited in the literature on non-stationary processes.

During the thirty five years at the Andhra University, he published forty papers in the field of Mathematics, Mathematical Physics and Statistics in journals of national and international repute, like: the Journal of London Math. Society, Annals of Mathl. Statistics, *Sankhya*, the Cal. Stat. Association's Bulletin, *Econometrica*, Canadian Journal of Maths. and so on. He was also a Visiting Professor at the Indian Statistical Institute (1954). He was invited to deliver lectures at the University of Travancore (1954) on 'Stochastic Processes', at Oak Ridge National Laboratory of U.S. Atomic Commission (1966) on 'Linear Prediction in Time Series', at University of Mysore (1968) on 'Non-stationary Processes' and at Indian Statistical Institute (1969) on 'Spectrum of Non-Stationary Processes and Mean Ergodic Theorem'. He authored two books, one 'The primary process of a smoothing relation' (his doctoral thesis published in 1951 in *Arkiv.f. Mat.* of the Royal Akad. of Sweden) and another, jointly with Prof. C. R. Rao, in the preparation of International Statistical Institute's bibliography in 'Time Series and Stochastic Processes' (published by Oliver Boyd in 1965 and edited by Prof. Herman Wold). He conducted several surveys and submitted expert reports on urbanisation of Visakhapatnam, notable among these being the Socio-economic Survey Reports of the Nagarjuna Sagar Project (in 4 volumes jointly with Prof. B. Sarveswara Rao in 1970). He was a member of the Review Committee for Statistics and Expert Committee for the Revision of Consumer Price Index, member of Research Advisory Committee of the Indian Council of Agricultural Research, Editorial Committee member of the Statistics Section of Telugu Encyclopaedia, Technical Committee member for Farm Management Studies in West Godavari District, Steering Committee member of Indian Econometric Conference, member of Research Advisory Committee of '*ATIRA*', member of the Editorial Committee of '*Sankhya*' and member of the Visiting Committee of the University Grants Commission. He was elected President of the Statistics Section of the Indian Science Congress in 1956. He was also the founder member of the Andhra Pradesh Academy of Sciences.

Although a gifted teacher and an eminent scholar of international repute, he did not shirk other duties. He still found time to evince keen interests in the administration of the Department in particular and the University in general. He was Warden, University Hostels; Secretary, Andhra University Information Bureau for studies outside India, and

Secretary, Old Students Association. He was an excellent sportsman, very good at playing Tennis, Badminton and Cards (Bridge game).

Prof. Nagabhushanam had built up the Department of Statistics on extremely sound lines with his characteristic commitment to high moral and academic values. These efforts quite naturally brought laurels to the department. Over the years, it produced several worthy students, most of whom, later as Professors, exemplarily shaped the destinies of many departments of statistics and mathematics not only in India but abroad as well. For over 35 years, Prof. Nagabhushanam served the University and the cause of higher education with ability and distinction in many capacities. Easily approachable, he endeared himself to one and all. He was a dedicated teacher, a sound scholar, a generous critic and above all a perfect gentleman. He was one of those rare teachers who command affection and admiration of generations of students and loyalty of friends and colleagues. Always elegantly dressed, cigar was his constant companion. Beneath the quiet and dignified exterior, lay immense warmth for his students and a great reservoir of energy for hard work. Not many, however, knew about the ever smiling and cheerful Professor's worries and disappointments. These he kept strictly to himself. He passed away on 15 February, 1979, at Secunderabad (India) having completed 71 years of age, leaving behind his wife, a son and two daughters and, of course, a host of friends, students and admirers, who still cherish their association with him.

N. VENUGOPAL

Preface

Time is the best barometer to gauge the worth of a person and his works who is no more in the mortal world. The commemorative acts and other memorials to perpetuate the glory of the departed soul stand out as monuments in time infinitum. These proclaim the values for which the person had stood in his life.

The present commemorative volume is one such epochal landmark in the field of statistics and is meant to perpetuate the memory of the towering personality of Professor Kandula Nagabhushanam (1908–1979).

That this volume is being brought out after a little more than a decade since his demise, brings out the fact with all emphasis that his reputation has stood the test of time.

The idea of bringing out the present volume emanated from the humble efforts of a small group of academics at the Andhra University, Waltair, in the early eighties, soon after Professor K. Nagabhushanam's death.

This group, in right earnest, began with making an appeal to K.N.'s admirers, students and friends to raise funds to do some useful academic deeds to perpetuate the memory of the departed soul. The response was spontaneous and stunningly large. Encouraged by this, Professor K. Nagabhushanam Memorial Society was got registered. It has Professor B. Sarveswara Rao (President) and Sri K. D. Rama Rao (Secretary and Treasurer). Other Executive Committee members are Professors: J. Gopalakrishna, M. Jagadeswara Rao, C. S. K. Bhagawan (now dead), K. V. N. Sarma (Late KN's son), K. V. Ramana, Dr. Y. R. K. Sarma, Dr. P. Samba Murthy and Sri K. Narasimham. The Society initiated several academic activities like starting a statistics library at the Department of Statistics, Andhra University, instituting a number of scholarships and prizes for poor and meritorious students and arranging Professor K.N.'s memorial lectures.

Later, in 1982, a new dimension was given to the activities of the Society, namely the bringing out of a 'Commemorative Volume' of international standard. The idea was mine but it was Sri Rama Rao who with all seriousness and zeal proceeded to finalise the matter with the Society.

Professor K.N.'s Commemoration Volume's Editorial Board was formed with Professor C. R. Rao (Chief Editor), Professor N. Venugopal (Editorial Secretary) and Professors: B.L.S. Prakasa Rao, B. R. Bhat, C. G. Khatri, S. R. Adke and J. Gopalakrishna as the other members. The project involved a lot of planning, execution and related effort. With the editorial committee members scattered all over India as well as abroad, the initial pace of the work was rather sluggish.

However, due to the excellent and willing co-operation of all the editorial committee members and friends, the project reached successful culmination. All the contributed articles that comprise this volume are invited ones from eminent scholars in the fields of their specialisation pertaining to stochastic processes, statistical inference and econometrics, the three areas dear to the late Professor.

With a sense of fulfillment, extreme delight and deep gratitude, I mention Mr. H. S. Poplai of Wiley Eastern, Delhi for his sustained interest; Professor C. R. Rao for giving us all support, lending his good name and agreeing to write the foreword (thus raising the stature of the volume), in addition to his own personal contribution to the volume; Professor Prakasa Rao, Bhat and Khatri (now dead) for their excellent help and counselling, Professor Adke and Gopalakrishna for the assistance provided, and all the contributors who readily accepted our invitation to contribute to this volume.

Our thanks are also due to all the learned academics who so kindly acted as referees.

Finally, the Professor K. N. Memorial Society (Waltair) deserves commendation for sponsoring the project 'Contributions to Stochastics'.

24 March, 1992 N. VENUGOPAL

Contents

1

Linear Transformations, Projection Operators and Generalized Inverses: A Geometric Approach*

C. Radhakrishna Rao

1.1. INTRODUCTION

A generalized inverse of a linear transformation $\mathbf{A}: v \rightarrow w$, where v and w are finite dimensional vector spaces, is defined using geometric concepts of linear transformations and projection operators. The inverse is uniquely defined in terms of specified subspaces $m \subset v$, $l \subset w$ and a linear transformation $\mathbf{N}$ such that $\mathbf{AN} = \mathbf{O}$, $\mathbf{NA} = \mathbf{O}$. Such an inverse which is unique is called the lm N-inverse. A Moore-Penrose type inverse is obtained by putting $\mathbf{N} = \mathbf{O}$. Applications to optimization problems when v and w are inner product spaces, such as least squares in a general setting, are discussed here. The results given in the paper can be extended without any major modification of proofs to bounded linear operators with closed range on Hilbert spaces.

Let v and w be two finite dimensional vector spaces and $\mathbf{A}: v \rightarrow w$ be a linear transformation from v to w. We denote the range of $\mathbf{A}$ by

$$\mathbf{R}(\mathbf{A}) = \{\mathbf{y} : \mathbf{Ax} = \mathbf{y},\ \mathbf{x} \in v\} = a \subset w \tag{1.1}$$

and the nullity or kernel of $\mathbf{A}$ by

$$\mathbf{K}(\mathbf{A}) = \{\mathbf{x} \in v : \mathbf{Ax} = 0\} = k \subset v. \tag{1.2}$$

If c and d are subspaces of v such that

$$c \cap d = 0, \text{ the null vector} \tag{1.3}$$

$$c \oplus d = \{\mathbf{x}_1 + \mathbf{x}_2 : \mathbf{x}_1 \in c,\ \mathbf{x}_2 \in d\} = v \tag{1.4}$$

then c and d are said to be direct complements. Given any subspace $c \subset v$, then there exists a subspace $d \subset v$, called the direct complement

*Research sponsored by the Air Force Office of Scientific Research under Grant AFSO-88-0030. The United States Government is authorized to reproduce and distribute reprints for governmental purposes notwithstanding any copyright notation hereon.

of c, which satisfies the conditions (1.3) and (1.4). The choice of d is not unique. Given any pair c, d satisfying (1.3) and (1.4) any vector $\mathbf{x} \in v$ has the unique decomposition

$$\mathbf{x} = \mathbf{x}_1 + \mathbf{x}_2, \mathbf{x}_1 \in c, \mathbf{x}_2 \in d. \tag{1.5}$$

The transformation $\mathbf{x} \to \mathbf{x}_1$, which is linear, is called the projection operator on c along d and is represented by

$$\mathbf{P}_{c \cdot d} : v \to v. \tag{1.6}$$

It is seen that

$$\mathbf{P}_{c \cdot d} + \mathbf{P}_{d \cdot c} = \mathbf{I}, \mathbf{P}_{c \cdot d}\mathbf{P}_{d \cdot c} = \mathbf{O} \tag{1.7}$$

$$\mathbf{R}(\mathbf{P}_{c \cdot d}) = c, \mathbf{R}(\mathbf{P}_{d \cdot c}) = d \tag{1.8}$$

where $\mathbf{I}$ and $\mathbf{O}$ are identity and null transformations on v.

$\mathbf{A} : v \to w$ is called bijective if it is onto and one-to-one. In such a case, there is an inverse of transformation $\mathbf{A}$, denoted by $\mathbf{A}^{-1}$, which is linear and is such that

$$\mathbf{A}\mathbf{A}^{-1} = \mathbf{I}\,(\text{on } w), \mathbf{A}^{-1}\mathbf{A} = \mathbf{I}\,(\text{on } v). \tag{1.9}$$

The restriction of $\mathbf{A} : v \to w$ to a subspace $c \subset v$ is also a linear transformation and is denoted by

$$\mathbf{A} \mid c : c \to w. \tag{1.10}$$

Let k be the nullity of $\mathbf{A}$ as defined in (1.2), m be any direct complement of k, (i.e. $k \oplus m = v$), and a be the range of $\mathbf{A}$. Then

$$\mathbf{A} \mid m : m \to a \text{ is a bijection.} \tag{1.11}$$

This is easily seen as follows. Choose $\mathbf{y} \in a$ and let $\mathbf{x} \in v$ be a preimage of $\mathbf{y}$, i.e., $\mathbf{x}$ is such that $\mathbf{Ax} = \mathbf{y}$. Then

$$\mathbf{P}_{m \cdot k}\mathbf{x} \in m \text{ and } \mathbf{Ax} = \mathbf{A}(\mathbf{P}_{m \cdot k} + \mathbf{P}_{k \cdot m}\mathbf{x}) = \mathbf{AP}_{m \cdot k}\mathbf{x} = \mathbf{y}$$

so that the mapping (1.11) is onto. Further, if $\mathbf{x}_1 \in m$ and $\mathbf{x}_2 \in m$, then $\mathbf{x}_1 - \mathbf{x}_2 \in m$. If, now $\mathbf{Ax}_1 = \mathbf{Ax}_2$, then $\mathbf{A}(\mathbf{x}_1 - \mathbf{x}_2) = 0 \Rightarrow \mathbf{x}_1 - \mathbf{x}_2 \in k$ which is a contradiction unless $x_1 = x_2$. Thus $\mathbf{x}_1 \neq \mathbf{x}_2 \Rightarrow \mathbf{Ax}_1 \neq \mathbf{Ax}_2$, which shows that the mapping (1.11) is one to one.

The result (1.11) implies the existence of an inverse

$$(\mathbf{A} \mid m)^{-1} = \mathbf{T}_m : a \to m \tag{1.12}$$

such that

$$\mathbf{T}_m\mathbf{A} \mid m = \mathbf{I}\,(\text{on } m), \mathbf{AT}_m\mathbf{A} = \mathbf{A}, \mathbf{T}_m\mathbf{A} = \mathbf{P}_{m \cdot k} \tag{1.13}$$

$$\mathbf{AT}_m \mid a = \mathbf{I}\,(\text{on } a), \mathbf{T}_m\mathbf{AT}_m = \mathbf{T}_m. \tag{1.14}$$

Note that $\mathbf{T}_m$ is defined only on $a \subset w$ and not the whole of w.

In this paper, we develop the concept of a generalized inverse (g-inverse) of $\mathbf{A}$ when it is not necessarily bijective and provide some characterizations and applications. This is accomplished by extending $\mathbf{T}_m$ as a linear transformation on any complement of a in w.

1.2 *G*-INVERSE OF A LINEAR TRANSFORMATION

Let $\mathbf{A} : v \to w$ be a linear transformation with range $a \subset w$ and nullity $k \subset v$. Consider a consistent linear equation $\mathbf{Ax} = \mathbf{y}$ $(\in a)$ and denote the set of solutions by

$$\mathbf{S}_y = \{\mathbf{x} \in v : \mathbf{Ax} = y\}. \tag{2.1}$$

Suppose that there exists a linear transformation $\mathbf{G} : w \to v$ such that

$$\mathbf{Gy} \in \mathbf{S}_y \text{ for any } \mathbf{y} \in a \tag{2.2}$$

i.e., $\mathbf{Gy}$ is a solution of $\mathbf{Ax} = \mathbf{y}$ for any $\mathbf{y} \in a$. Then substituting $\mathbf{Gy}$ for $\mathbf{x}$ in $\mathbf{Ax} = \mathbf{y}$,

$$\mathbf{AGy} = \mathbf{y} \ \forall\ \mathbf{y} \in a \text{ or } \mathbf{AG} \mid a = \mathbf{I} \text{ (on } a) \Rightarrow \mathbf{AGA} = \mathbf{A}. \tag{2.3}$$

Conversely, if there exists a $\mathbf{G}$ such that (2.3) holds and $\mathbf{Ax} = \mathbf{y}$ is a consistent equation, i.e., there exists an $\mathbf{x}$ such that $\mathbf{Ax} = \mathbf{y}$, then

$$\mathbf{AGA} = \mathbf{A} \Rightarrow \mathbf{AGAx} = \mathbf{Ax} \text{ or } \mathbf{AGy} = \mathbf{y} \tag{2.4}$$

so that $\mathbf{Gy}$ is a solution of $\mathbf{Ax} = \mathbf{y}$. Any $\mathbf{G}$ satisfying the condition (2.3) is called a *g*-inverse of $\mathbf{A}$; this enables us to solve a consistent equation $\mathbf{Ax} = \mathbf{y}$. Such a $\mathbf{G}$ which may not be unique is denoted by $\mathbf{A}^-$ and the set of all $\mathbf{G}$ satisfying (2.3) is denoted by $\{\mathbf{A}^-\}$. The following lemmas characterize *g*-inverses.

Lemma 2.1. Let k be the nullity of $\mathbf{A}$, a the range of $\mathbf{A}$, and $\mathbf{G}$ be a linear transformation such that $\mathbf{AGA} = \mathbf{A}$, i.e., $\mathbf{G} \in \{\mathbf{A}^-\}$. Then the following holds:

(i) $m = \mathbf{R}(\mathbf{GA})$ is a direct complement of k, and $\mathbf{T}_m : a \to m = \mathbf{G} \mid a$ or $\mathbf{GA} \mid m = \mathbf{I}$ (on m), where $\mathbf{T}_m = (\mathbf{A} \mid m)^{-1}$.

(ii) If $\mathbf{N} = \mathbf{G} - \mathbf{GAG}$, then $\mathbf{AN} = \mathbf{O}$ and $\mathbf{NA} = \mathbf{O}$.

(iii) $l = \mathbf{R}(\mathbf{I} - \mathbf{AG})$ is a direct complement of a, and $\mathbf{N} \mid l = \mathbf{G} \mid l$.

Proof of (i). Let $\mathbf{x} \in m \cap k$. Then $\mathbf{x} = \mathbf{GA}u$ for some u and $\mathbf{Ax} = \mathbf{AGA}u = \mathbf{Au} = 0 \Rightarrow \mathbf{x} = 0$, i.e., $m \cap k = 0$. Also, if $\mathbf{x} \in v$, then

$$\mathbf{x} = \mathbf{GAx} + (\mathbf{I} - \mathbf{GA})\mathbf{x} = \mathbf{x}_1 + \mathbf{x}_2$$

where $\mathbf{x}_1 \in m$ and $\mathbf{x}_2 \in k$ since $\mathbf{Ax}_2 = 0$. Thus the conditions (1.3) and (1.4) are satisfied so that m is a direct complement of k. Then there exists $\mathbf{T}_m$ as defined in (1.12)-(1.14), and in particular $\mathbf{T}_m\mathbf{A} \mid m = \mathbf{I}$. Now

$$\mathbf{T}_m\mathbf{Ax} = \mathbf{T}_m\mathbf{AGAx} = \mathbf{T}_m\mathbf{A}(\mathbf{GAx}) = \mathbf{GAx}$$

since $\mathbf{GAx} \in m$, which implies $\mathbf{T}_m = \mathbf{G} \mid a$ or $\mathbf{GA} \mid m = \mathbf{I}$.

Proof of (ii). Obvious.

Proof of (iii). Let $\mathbf{y} \in l \cap a$. Then, for some $\mathbf{x}$ and $\mathbf{u}$,

$$\mathbf{y} = \mathbf{Ax} = (\mathbf{I} - \mathbf{AG})\mathbf{u} \Rightarrow \mathbf{AGAx} = \mathbf{AG}(\mathbf{I} - \mathbf{AG})\mathbf{u} = 0 \Rightarrow \mathbf{Ax} = 0$$

so that $l \cap a = 0$. Further

$$\mathbf{y} = \mathbf{AGy} + (\mathbf{I} - \mathbf{AG})\mathbf{y} = \mathbf{y} = \mathbf{y}_1 + \mathbf{y}_2$$

where $\mathbf{y}_1 \in a$ and $\mathbf{y}_2 \in l$. Thus the conditions (1.3) and (1.4) are satisfied so that l is a direct complement of a. Consider

$$\mathbf{N}(\mathbf{I} - \mathbf{AG})\mathbf{u} = \mathbf{Nu} = (\mathbf{G} - \mathbf{GAG})\mathbf{u} = \mathbf{G}(\mathbf{I} - \mathbf{AG})\mathbf{u}$$

for any $\mathbf{u}$, so that $\mathbf{N} \mid l = \mathbf{G} \mid l$. Lemma 1 is proved.

Definition. Let m be any direct complement of k (the nullity of **A**) and l be any direct complement of a (the range of **A**) and $\mathbf{N} : w \to v$ be any linear transformation such that $\mathbf{AN} = \mathbf{O}$ and $\mathbf{NA} = \mathbf{O}$. Then **G** is said to be an $l m$**N**-inverse of **A**, if

$$\text{(i) } \mathbf{G} \text{ is linear, (ii) } \mathbf{GA} \mid m = \mathbf{I} \text{ and (iii) } \mathbf{G} \mid l = \mathbf{N} \mid l. \tag{2.5}$$

Such a **G**, if it exists, is denoted by $\mathbf{G}_{lm\mathbf{N}}$. Lemma 2.2 characterizes such inverses.

Lemma 2.2.

$$\mathbf{G}_{lm\mathbf{N}} = \mathbf{T}_m \mathbf{P}_{a \cdot l} + \mathbf{N}, \tag{2.6}$$

where $\mathbf{T}_m$ is as defined in (1.12) with respect to a chosen m, so that $\mathbf{G}_{lm\mathbf{N}}$ is unique for any given l, m and **N**.

Proof. Let $\mathbf{y} = \mathbf{y}_1 + \mathbf{y}_2$, $\mathbf{y}_1 \in a$ and $\mathbf{y}_2 \in l$. Then

$$\begin{aligned} \mathbf{G}_{lm\mathbf{N}}\mathbf{y} &= \mathbf{G}_{lm\mathbf{N}}\mathbf{y}_1 + \mathbf{G}_{lm\mathbf{N}}\mathbf{y}_2 \\ &= \mathbf{T}_m\mathbf{y}_1 + \mathbf{N}\mathbf{y}_2, \text{ using (2.5)} \\ &= \mathbf{T}_m\mathbf{P}_{a \cdot l}\mathbf{y} + \mathbf{N}\mathbf{y} \\ &= (\mathbf{T}_m\mathbf{P}_{a \cdot l} + \mathbf{N})\mathbf{y} \;\forall\; \mathbf{y} \in w, \end{aligned}$$

which establishes Lemma 2.2.

Lemma 2.3 $\mathbf{G}_{lm\mathbf{N}} \in \{\mathbf{A}^-\}$ and the mapping

$$(l, m, \mathbf{N}) \to \{\mathbf{A}^-\} \text{ is bijective.} \tag{2.7}$$

Proof. Using the representation (2.6).

$$\begin{aligned} \mathbf{AG}_{lm\mathbf{N}}\mathbf{A} &= \mathbf{A}(\mathbf{T}_m\mathbf{P}_{a \cdot l} + \mathbf{N})\mathbf{A} \\ &= \mathbf{AT}_m\mathbf{P}_{a \cdot l}\mathbf{A} = \mathbf{AT}_m\mathbf{A} = \mathbf{A}, \text{ using (1.13)} \end{aligned}$$

so that $\mathbf{G}_{lm\mathbf{N}} \in \{\mathbf{A}^-\}$.

Again the representation (2.6) shows that $\mathbf{G}_{lm\mathbf{N}}$ is unique for given l, m, **N**. On the other hand, it has been shown in Lemma 2.1 that given a $\mathbf{G} \in \{\mathbf{A}^-\}$, there exists an l, m, **N** for which it is the lm**N**-inverse, and the result (2.7) is established.

Incidentally, the existence of a g-inverse **G** satisfying the condition $\mathbf{AGA} = \mathbf{A}$ is established through the representation (2.6).

The anatomy of a g-inverse is exhibited in the following figure.

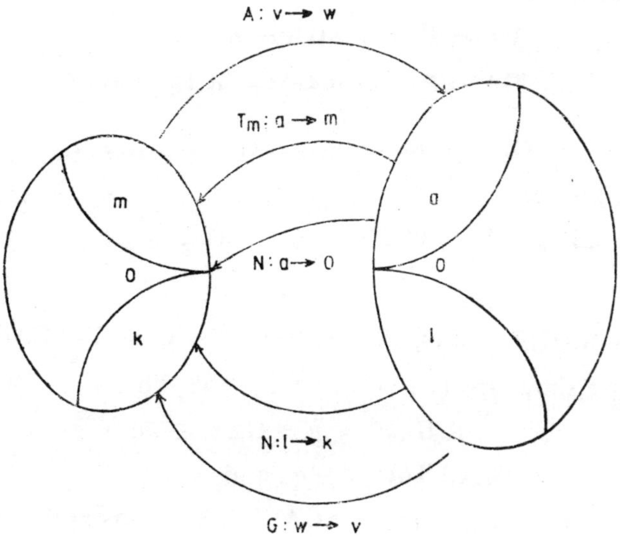

Fig. 1.1. Diagram representing *g*-inverse.

Lemma 2.4. Let k, m, a, l and $\mathbf{N}$ be as given in the definition above the equation (2.5). Then the following statements on a linear transformation $\mathbf{G}$ are equivalent.

(i) $\mathbf{GA} \mid m = \mathbf{I}$, $\mathbf{G} \mid l = \mathbf{N} \mid l$ (as defined in (2.5)).

(ii) $\mathbf{GA} = \mathbf{P}_{m \cdot k}$, $\mathbf{GP}_{l \cdot a} = \mathbf{N}$.

(iii) $\mathbf{GA} = \mathbf{P}_{m \cdot k}$, $\mathbf{AG} = \mathbf{P}_{a \cdot l}$, $\mathbf{P}_{k, m}\mathbf{G} = \mathbf{N}$.

(iv) $\mathbf{GA} = \mathbf{P}_{m \cdot k}$, $\mathbf{AG} = \mathbf{P}_{a \cdot l}$, $\mathbf{G} - \mathbf{GAG} = \mathbf{N}$.

(v) $\mathbf{AGA} = \mathbf{A}$, $\mathbf{R}(\mathbf{G} \mid a) = m$, $\mathbf{GP}_{l \cdot a} = \mathbf{N}$.

(vi) $\mathbf{AGA} = \mathbf{A}$, $\mathbf{R}(\mathbf{G} \mid \mathbf{A}) = m$, $\mathbf{R}(\mathbf{G} \mid l) \subset k$, $\mathbf{G} - \mathbf{GAG} = \mathbf{N}$.

Proof. (i) ⇒ (ii).

$$\mathbf{GAx} = \mathbf{x} \text{ if } \mathbf{x} \in m, \text{ using the condition } \mathbf{GA} \mid m = \mathbf{I},$$

$$\mathbf{GAx} = 0 \text{ if } \mathbf{x} \in k, \text{ using the condition } \mathbf{Ax} = 0$$

which shows that $\mathbf{GA} = \mathbf{P}_{m \cdot k}$. Further

$$\mathbf{G} \mid l = \mathbf{N} \mid l \Rightarrow \mathbf{GP}_{l \cdot a} = \mathbf{NP}_{l \cdot a} = \mathbf{NP}_{l \cdot a} + \mathbf{NP}_{a \cdot l} = \mathbf{N}.$$

(ii) ⇒ (iii).

$$\mathbf{GP}_{l \cdot a} = \mathbf{N} \Rightarrow \mathbf{AGP}_{l \cdot a} = \mathbf{0} \text{ and } \mathbf{A} = \mathbf{AP}_{m \cdot k} = \mathbf{AGA}$$

which together imply $\mathbf{AG} = \mathbf{P}_{a \cdot l}$. Further

$$\mathbf{N} = \mathbf{GP}_{l \cdot a} = \mathbf{G}(\mathbf{I} - \mathbf{P}_{a \cdot l}) = \mathbf{G}(\mathbf{I} - \mathbf{AG})$$
$$= (\mathbf{I} - \mathbf{GA})\mathbf{G} = (\mathbf{I} - \mathbf{P}_{m \cdot k})\mathbf{G} = \mathbf{P}_{k \cdot m}\mathbf{G}.$$

(iii) ⇒ (iv), since

$$\mathbf{P}_{k \cdot m} = \mathbf{I} - \mathbf{GA} \text{ and } \mathbf{P}_{k \cdot m}\mathbf{G} = \mathbf{G} - \mathbf{GAG}.$$

(iv) ⇒ (v), since

$$GA = P_{m\cdot k} \Rightarrow AGA = AP_{m\cdot k} = A,$$

$$GA = P_{m\cdot k} \Rightarrow R(GA) = R(P_{m\cdot k}) = m$$

and

$$GP_{l\cdot a} = G - GP_{a\cdot l} = G - GAG = N.$$

(v) ⇒ (vi), since

$$GP_{l\cdot a} = N \Rightarrow AGP_{l\cdot a} = AN = O \Rightarrow R(GP_{l\cdot a}) \subset k.$$

Further,

$$(G - GAG)P_{l\cdot a} = N,\ (G - GAG)P_{a\cdot l} = O \Rightarrow G - GAG = N.$$

Finally we show that (vi) ⇒ (i). Let $x_1 \in m$. Then

$$R(G \mid a) = m \Rightarrow GAx_1 \in m.$$

Now let $x_1 + s = GAx_1$. Then $s \in m$, and

$$Ax_1 + As = A(x_1 + s) = AGAx_1 = Ax_1 \Rightarrow As = 0 \Rightarrow s \in k$$

using the condition $AGA = A$. Since m and k have only the null vector as intersection, $s = 0$, which shows that $GAx_1 = x_1$ or $GA \mid m = I$. Further noting that

$$R(G \mid l) \subset k \Rightarrow AGP_{l\cdot a} = 0$$

we have

$$N = G - GAG \Rightarrow NP_{l\cdot a} = GP_{l\cdot a} - GAGP_{l\cdot a} = GP_{l\cdot a}$$

$$\Rightarrow N \mid l = G \mid l.$$

Lemma 2.4 is established.

When $N = O$, the statement (iv) of Lemma 2.4 reduces to the definition of a *g*-inverse given by Nashed and Voturba (1976), so that their *g*-inverse is G_{lmO} which is discussed in the next section.

1.3. REFLEXIVE *G*-INVERSE

Definition. G is said to be a reflexive (or an outer inverse) of A if

$$AGA = A \text{ and } G = GAG. \tag{3.1}$$

The class of G satisfying (3.1) is represented by $\{A_r^-\}$ and any member of it by A_r^-.

The following lemma characterizes reflexive *g*-inverses:

Lemma 3.1.

$$\{G_{lmO}\} = \{A_r^-\} \subset \{A^-\} \tag{3.2}$$

i.e., all reflexive *g*-inverses are generated by choosing $N = O$, and $\{l, m\}$ as the set of all direct complements of a and k respectively.

Proof. From the representation (2.6),

$$G_{lmO} = T_m P_{a\cdot l}.$$

Then.

$$\mathbf{AT}_m\mathbf{P}_{a.l}\mathbf{A} = \mathbf{AT}_m\mathbf{A} = \mathbf{A}, \text{ using (1.13)}$$

$$\mathbf{T}_m\mathbf{P}_{a.l}\mathbf{AT}_m\mathbf{P}_{a\cdot l} = \mathbf{T}_m\mathbf{AT}_m\mathbf{P}_{a.l} = \mathbf{T}_m\mathbf{P}_{a.l}, \text{ using (1.14)}$$

which shows that $\mathbf{G}_{lm\mathbf{O}} \in \{\mathbf{A}_r^-\}$.

Conversely let $\mathbf{AGA} = \mathbf{A}$ and $\mathbf{G} = \mathbf{GAG}$. Choose $m = \mathbf{R}(\mathbf{GA})$ and $l = \mathbf{R}(\mathbf{I} - \mathbf{AG})$. Then as in (i) of Lemma 2.1,

$$\mathbf{G} \mid a = \mathbf{T}_m.$$

Further if $\mathbf{y} \in l$, then $\mathbf{y} = (\mathbf{I} - \mathbf{AG})\mathbf{u}$ for some $\mathbf{u}$ so that

$$\mathbf{G}y = \mathbf{G}(\mathbf{I} - \mathbf{AG})\mathbf{u} = \mathbf{0}, \text{ since } \mathbf{G} = \mathbf{GAG}$$
$$\Rightarrow \mathbf{G} \mid l = 0.$$

Thus $\mathbf{G}$ satisfies the definition of $\mathbf{G}_{lm\mathbf{O}}$ and Lemma 3.1 is proved.

Lemma 3.2

$$\mathbf{G}_{lm\mathbf{N}} = \mathbf{G}_{lm\mathbf{O}} + N \tag{3.3}$$

Proof. The result (3.3) follows from the representation (2.6).

Lemma 3.3. The following statements are equivalent.

(i) $\mathbf{G}$ is the $lm\mathbf{O}$-inverse.

(ii) $\mathbf{GA} = \mathbf{P}_{m.k}$, $\mathbf{G} \mid l = \mathbf{O}$.

(iii) $\mathbf{GA} = \mathbf{P}_{m.k}$, $\mathbf{AG} = \mathbf{P}_{a\cdot l}$, $\mathbf{G} = \mathbf{GAG}$.

(iv) $\mathbf{GA} = \mathbf{P}_{m\cdot k}$, $g = \mathbf{R}(\mathbf{G}) = m$, $\mathbf{AG} = \mathbf{P}_{a\cdot l}$.

(v) $\mathbf{AGA} = \mathbf{A}$, $g = \mathbf{R}(\mathbf{G}) = m$, $\mathbf{R}(\mathbf{G} \mid l) = 0$.

The results are proved on the same lines as in Lemma 2.4.

If v and w are inner product spaces and if we choose m as the orthogonal complement of k, and l as that of a, then the condition (iv) of Lemma 3.3 can be written as

$$\mathbf{GA} = \mathbf{I} - \mathbf{P}_k = \mathbf{P}_g, \ \mathbf{AG} = \mathbf{P}_a \tag{3.4}$$

where $g = \mathbf{R}(\mathbf{G})$, $\mathbf{P}_k$ is the orthogonal projector on k and $\mathbf{P}_a$ is the orthogonal projector on a, which is equivalent to the definition given by Moore (1920) and Penrose (1955).

1.4. *LM*-INVERSE

If in the definition of the lm $\mathbf{N}$-inverse, we do not specify $\mathbf{N}$ but only require that $\mathbf{G} \mid l \subset k$, then we can write the conditions for a g-inverse in the form is

$$\mathbf{GA} \mid m = \mathbf{I}, \ \mathbf{AG} \mid l = \mathbf{O}. \tag{4.1}$$

We represent the solution of (4.1) by $\mathbf{G}_{lm}$, which may not be unique and call it an lm-inverse.

Lemma 4.1. The following statements are equivalent for given l and m, any direct complements of a and k respectively.

(i) $\mathbf{G}$ is an lm-inverse.

(ii) $\mathbf{GA} = \mathbf{P}_{m \cdot k}$, $\mathbf{AG} = \mathbf{P}_{a \cdot l}$.
(iii) $\mathbf{AGA} = \mathbf{A}$, $\mathbf{R}(\mathbf{G} \mid a) = m$, $\mathbf{R}(\mathbf{G} \mid l) \subset k$.

The proof of Lemma 4.1 follows on the same lines as that of Lemma 2.4.

The definition given in (ii) of Lemma 4.1 was proposed by Langehop (1967).

1.5. OTHER INVERSES

1.5.1. *M*-inverse

An *m*-inverse is defined by specifying *m* only as in the following lemma.

Lemma 5.1. The following conditions are equivalent for an *m*-inverse:
(i) $\mathbf{GA} \mid m = \mathbf{I}$.
(ii) $\mathbf{GA} = \mathbf{P}_{m \cdot k}$.
(iii) $\mathbf{AGA} = \mathbf{A}$, $\mathbf{R}(\mathbf{G} \mid a) = m$.

We represent an *m*-inverse of **A**, which may not be unique, by $\mathbf{A}_m^-$ to be consistent with the notation developed earlier by the author (Rao, 1967).

If *v* is a vector space endowed with an inner product, then we may choose *m* to be an orthogonal complement of *k*. In such a case, if $\mathbf{Ax} = \mathbf{y}$ is a consistent equation, then

$$\min_{\mathbf{Ax}=\mathbf{y}} \|\mathbf{x}\| = \|\mathbf{A}_m^- \mathbf{y}\| \tag{5.1}$$

so that $\mathbf{A}_m^- \mathbf{y}$ is the minimum norm solution of $\mathbf{Ax} = \mathbf{y}$. Note that any particular solution of $\mathbf{Ax} = \mathbf{y}$ is given by $\mathbf{A}^- \mathbf{y}$ where $\mathbf{A}^-$ is any *g*-inverse satisfying the condition $\mathbf{AA}^-\mathbf{A} = \mathbf{A}$.

To prove 1.5.1, we observe that if **x** is any solution of $\mathbf{Ax} = \mathbf{y}$, then

$$\mathbf{A}(\mathbf{x} - \mathbf{A}_m^- \mathbf{y}) = 0 \Rightarrow x - \mathbf{A}_m^- \mathbf{y} \in k$$

and by definition $\mathbf{A}_m^- \mathbf{y} \in m$, and since *m* and *k* are orthogonal complements,

$$\begin{aligned} \|\mathbf{x}\|^2 &= \|\mathbf{x} - \mathbf{A}_m^- \mathbf{y} + \mathbf{A}_m^- \mathbf{y}\|^2 \\ &= \|\mathbf{x} - \mathbf{A}_m^- \mathbf{y}\|^2 + \|\mathbf{A}_m^- \mathbf{y}\|^2, \text{ by Pythagorous theorem} \\ &\geqslant \|\mathbf{A}_m^- \mathbf{y}\|^2. \end{aligned}$$

1.5.2. *l*-inverse

An *l*-inverse of **A**, denoted by $\mathbf{A}_l^-$ is a linear transformation **G** as characterized in Lemma 5.2.

Lemma 5.2. The following conditions are equivalent for an *l*-inverse.
(i) $\mathbf{AG} = \mathbf{P}_{a \cdot l}$.
(ii) $\mathbf{AGA} = \mathbf{A}$, $\mathbf{R}(\mathbf{G} \mid l) \subset k$.

If *w* is an inner product space, we may choose *l* to be the orthogonal complement of *a*. In such a case

$$\min_{x} \|\mathbf{y} - \mathbf{Ax}\| = \|\mathbf{y} - \mathbf{AA}_l^- \mathbf{y}\| \tag{5.2}$$

so that $\mathbf{A}_l^- \mathbf{y}$ is a general least squares solution of a possibly inconsistent equation. Observe that

$$\mathbf{y} - \mathbf{A}\mathbf{A}_l^- \mathbf{y} = (\mathbf{I} - \mathbf{P}_{a \cdot l})\mathbf{y} = \mathbf{P}_{l \cdot a}\mathbf{y} \in l$$

$$\mathbf{A}\mathbf{A}_l^- \mathbf{y} = \mathbf{P}_{a \cdot l}\mathbf{y} \in a \Rightarrow \mathbf{A}\mathbf{A}_l^- \mathbf{y} - \mathbf{A}\mathbf{x} \in a$$

so that by virtue of orthogonality of l and a

$$\begin{aligned} \|\mathbf{y} - \mathbf{A}\mathbf{x}\|^2 &= \|\mathbf{y} - \mathbf{A}\mathbf{A}_l^- \mathbf{y} + \mathbf{A}\mathbf{A}_l^- \mathbf{y} - \mathbf{A}\mathbf{x}\|^2 \\ &= \|\mathbf{y} - \mathbf{A}\mathbf{A}_l^- \mathbf{y}\|^2 + \|\mathbf{A}\mathbf{A}_l^- \mathbf{y} - \mathbf{A}\mathbf{x}\|^2 \geqslant \|\mathbf{y} - \mathbf{A}\mathbf{A}_l^- \mathbf{y}\|^2. \end{aligned}$$

1.5.3. *lm*O-inverse

The *lm***O**-inverse, specified by l, m, and $\mathbf{N} = \mathbf{O}$ and denoted by $\mathbf{A}_{lm\mathbf{O}}^-$, is already characterized in Lemma 5.3. When v and w are inner product spaces, we may choose m to be the orthogonal complement of k, and l to be the orthogonal complement of a. In such a case the *lm***O**-inverse can be characterized by

$$\mathbf{G}\mathbf{A} = (\mathbf{I} - \mathbf{P}_k) = \mathbf{P}_g, \ \mathbf{A}\mathbf{G} = \mathbf{P}_a \tag{5.3}$$

where $\mathbf{P}_x$ denotes the orthogonal projection operator on x. The inverse satisfying the conditions (5.3) is the Moore-Penrose inverse and is represented by $\mathbf{A}^+$. We show that

$$\min_{x_l} \{\|\mathbf{x}_l\| : \|\mathbf{y} - \mathbf{A}\mathbf{x}_l\| = \min_{x} \|\mathbf{y} - \mathbf{A}\mathbf{x}\|\} = \|\mathbf{A}^+\mathbf{y}\| \tag{5.4}$$

where the norms associated with $\mathbf{x} \in v$ and $\mathbf{y} \in w$ may be different, so that $\mathbf{A}^+\mathbf{y}$ is the minimum norm least squares solution of $\mathbf{A}\mathbf{x} = \mathbf{y}$ (possibly inconsistent equation).

Observe that

$$\|\mathbf{y} - \mathbf{A}\mathbf{x}\|^2 - \|\mathbf{P}_a\mathbf{y} - \mathbf{A}\mathbf{x}\|^2 + \|\mathbf{P}_l\mathbf{y}\|^2 \tag{5.5}$$

where the second member does not involve $\mathbf{x}$, and the minimum of $\|\mathbf{y} - \mathbf{A}\mathbf{x}\|$ is attained if $\mathbf{A}\mathbf{x} = \mathbf{P}_a\mathbf{y}$ so that $\mathbf{x}_l$ is a solution of $\mathbf{A}\mathbf{x} = \mathbf{P}_a\mathbf{y}$ which is a consistent equation. The problem (5.4) demands the solution of $\mathbf{A}\mathbf{x} = \mathbf{P}_a\mathbf{y}$ with the minimum norm, which is supplied by equation (5.1).

$$\hat{x} = \mathbf{A}_m^-\mathbf{P}_A\mathbf{y} = \mathbf{G}\mathbf{y}.$$

Now,

$$\mathbf{G}\mathbf{A} = \mathbf{A}_m^-\mathbf{P}_a\mathbf{A} = \mathbf{A}_m^-\mathbf{A} = \mathbf{P}_{m \cdot k} = (\mathbf{I} - \mathbf{P}_k)$$

$$\mathbf{A}\mathbf{G} = \mathbf{A}\mathbf{A}_m^-\mathbf{P}_a = \mathbf{P}_a, \text{ since } \mathbf{R}(\mathbf{P}_a) = a$$

so that $\mathbf{G}$ satisfies the conditions (5.3) and hence it is the *lm***O**-inverse.

Thus the *l*- and *lm***O**-inverses provide solutions to the least squares problem in the most general setting. When v and w are Euclidean spaces of m and n dimensions respectively, the linear transformation $\mathbf{A}$ can be represented as an $n \times m$ matrix in which case explicit representations can be obtained for various generalized inverses as shown in Rao (1955, 1962, 1967, 1973) and Rao and Mitra (1971). The geometric approach in the case of a general linear transformation was developed in Rao and Yanoi (1985). All the results of this paper can be extended without any major modification of the proofs to bounded linear operations with closed range on Hibert spaces.

REFERENCES

1. Langenhop, C. E. (1967): On generalized inverse of matrices. *Siam J. Appl. Math.* **15,** 1239–1246.

2. Moore, E. H. (1920): On the reciprocal of the general algebraic matrix (Abstract). *Bull. Amer. Math. Soc.* **26,** 394–395.

3. Nashed, M. Z. and Votruba, G. F. (1976): A unified operator theory of generalized inverses. In *Generalized Inverses and its Applications* (Ed. M. Z. Nashed), 1–110.

4. Penrose, R. (1955): A generalized inverse for matrices. *Proc. Cambridge Philos. Soc.,* **51,** 406–413.

5. Rao, C. Radhakrishna (1955): Analysis of dispersion for multiple classified data with unequal numbers in cells. *Sankhyā* **15,** 253–280.

6. ——— (1962): A note on the generalized inverse of a matrix with applications to problems of mathematical statistics. *J. Roy. Statist. Soc.* **B. 24,** 152–158.

7. ——— (1967): Calculus of generalized inverses of matrices: Part I—general theory. *Sankhyā,* **A 29,** 317–350.

8. ——— (1973): *Linear Statistical Inference and its Applications,* (Second Edition) Wiley, New York.

9. Rao, C. Radhakrishna and Mitra, S. K. (1971): *Generalized Inverse of Matrices and its Applications,* Wiley, New York.

10. Rao, C. Radhakrishna and Yanoi, H. (1985): Generalized inverse of linear transformations. *Linear Algebra and its Applications.* **66,** 87–98.

2

On Weighted Least Squares and Some Applications

B. L. S. Prakasa Rao

Indian Statistical Institute, New Delhi

2.1 INTRODUCTION

The concept of least squares for estimation of parameters plays a prominent role in linear models and non-linear regression problems. In linear models, it gives optimum estimators in some sense [cf. Rao (1973)] and in non-linear regression problems, it gives estimators which are asymptotically well-behaved [cf. Jennrich (1969), Prakasa Rao (1985)]. In autoregressive problems, the least squares approach gives estimators which are computable and which have nice asymptotic properties [cf. Basawa and Prakasa Rao (1980)]. However, the least squares approach for parameter estimation does not take into account any prior information about the parameter available to the statistician. It treats all the possible values of the parameter equally as far as the weightage in the computation of the least squares is concerned. However, suppose there is a prior information about the parameter θ (say) through its prior density (say) $\lambda^{-1}(\theta)$. It is more appropriate to give the weight $\lambda(\theta)$ to the error $Y_i - g_i(\theta)$ in the non-linear regression model $Y_i = g_i(\theta) + \epsilon_i$, $i \geqslant 1$ and compute the weighted least squares estimator (WLSE) by minimizing $\lambda(\theta) \sum_{i=1}^{n} (Y_i - g_i(\theta))^2$ rather than compute the least squares estimator (LSE) by minimizing $\sum_{i=1}^{n} (Y_i - g_i(\theta))^2$. Another instance where the weight function $\lambda(\theta)$ may be random, i.e., a function of the observations and the parameter θ when the error variance σ^2 is unknown. Then one may estimate σ^2 from the observations $Y_1, \ldots, Y_n$ (say) by $\hat{\sigma}^2(\theta)$. In such an event the parameter θ may be estimated by the least squares approach by minimising

$$Q_n(\theta) = \hat{\sigma}^2(\theta) \sum_{n=1}^{n} (Y_i - g_i(\theta))^2.$$

Even though the weighted least squares approach has been adopted by several authors earlier, it was mostly in connection with estimation of variance components. Observe that if $\lambda(\theta) \equiv 1$, the WLSE reduces to the ordinary LSE. We will give applications of this approach to linear models, non-linear regression models and autoregressive time series. It is shown that ridge estimators for linear models are special cases of WLSE for proper choice of $\lambda(\theta)$. Ridge estimators in non-linear regression models and autoregressive time series are derived here. Asymptotic properties of WLSE in non-linear regression models have been studied under some regularity conditions. It was shown that the WLSE are "better" than the LSE asymptotically in terms of asymptotic variance provided $\lambda'(\theta_0) = 0$ and $\lambda''(\theta_0) > 0$. The approach discussed here may be termed as a Bayesian type least squares method for the estimation problems.

2.2. LINEAR REGRESSION

Consider the linear regression model

$$\mathbf{Y} = \mathbf{X}\boldsymbol{\beta} + \boldsymbol{\epsilon} \tag{2.0}$$

where $\mathbf{Y}$ is $k \times l$ vector of observations, $\mathbf{X}$ is a design matrix of order $k \times m$, $\boldsymbol{\beta}$ is the vector $(\beta_1, \ldots, \beta_m)'$ of unknown parameters and $\boldsymbol{\epsilon}$ is a random vector with $E(\epsilon_i) = 0$, $1 \leqslant i \leqslant k$. Suppose $\lambda(\boldsymbol{\beta})$ is a given positive weight function (possibly depending on $\mathbf{X}$, $\mathbf{Y}$ and $\boldsymbol{\beta}$). $\hat{\boldsymbol{\beta}}$ is said to be a weighted least squares estimator (WLSE) of $\boldsymbol{\beta}$ if $\hat{\boldsymbol{\beta}}$ minimizes

$$\nu(\boldsymbol{\beta}) = \lambda(\boldsymbol{\beta})(\mathbf{Y} - \mathbf{X}\boldsymbol{\beta})'(\mathbf{Y} - \mathbf{X}\boldsymbol{\beta}) \tag{2.1}$$

For instance, $\boldsymbol{\beta}$ might have a prior density $\lambda^{-1}(\cdot)$ and one can interpret $\nu(\boldsymbol{\beta})$ as the sum of the squared errors with weights inversely proportional to the density $\lambda^{-1}(\boldsymbol{\beta})$. Note that the points with high density contribute less to $\nu(\boldsymbol{\beta})$ as compared with points with low density. If $\lambda(\boldsymbol{\beta}) \equiv 1$, then the function $\nu(\boldsymbol{\beta})$ reduces to the ordinary least squares.

Suppose that $\lambda(\boldsymbol{\beta})$ is differentiable with respect to components of $\boldsymbol{\beta}$. Then

$$\nabla\nu(\boldsymbol{\beta}) = \nabla\lambda(\boldsymbol{\beta})(\mathbf{Y} - \mathbf{X}\boldsymbol{\beta})'(\mathbf{Y} - \mathbf{X}\boldsymbol{\beta}) - 2\lambda(\boldsymbol{\beta})\mathbf{X}'(\mathbf{Y} - \mathbf{X}\boldsymbol{\beta}) \tag{2.2}$$

and $\hat{\boldsymbol{\beta}}$ is a solution of the equation $\nabla\nu(\boldsymbol{\beta}) = 0$. Hence

$$\nabla\lambda(\hat{\boldsymbol{\beta}})(\mathbf{Y} - \mathbf{X}\hat{\boldsymbol{\beta}})'(\mathbf{Y} - \mathbf{X}\hat{\boldsymbol{\beta}}) = 2\lambda(\hat{\boldsymbol{\beta}})(\mathbf{X}'\mathbf{Y} - \mathbf{X}'\mathbf{X}\hat{\boldsymbol{\beta}}) \tag{2.3}$$

where $\nabla\lambda(\hat{\boldsymbol{\beta}})$ denotes $\nabla\lambda(\boldsymbol{\beta})$ evaluated at $\boldsymbol{\beta} = \hat{\boldsymbol{\beta}}$. In case $\lambda(\boldsymbol{\beta}) \equiv 1$, then the equation (2.3) reduces to the standard normal equations. The equation (2.3) can be written in the form

$$\nabla \log \lambda(\hat{\boldsymbol{\beta}}) = 2(\mathbf{X}'\mathbf{Y} - \mathbf{X}'\mathbf{X}\hat{\boldsymbol{\beta}})[(\mathbf{Y} - \mathbf{X}\hat{\boldsymbol{\beta}})'(\mathbf{Y} - \mathbf{X}\hat{\boldsymbol{\beta}})]^{-1} \tag{2.4}$$

Suppose $\lambda(\boldsymbol{\beta})$ is given by the relation

$$\log \lambda(\boldsymbol{\beta}) = \lambda_1 + 2\boldsymbol{\lambda}_0'\boldsymbol{\beta} + \boldsymbol{\beta}'\mathbf{A}\boldsymbol{\beta}. \tag{2.5}$$

Then,

$$\nabla \log \lambda(\boldsymbol{\beta}) = 2\boldsymbol{\lambda}_0 + 2\mathbf{A}\boldsymbol{\beta}$$

and (2.4) reduces to

$$2\boldsymbol{\lambda}_0 + 2A\hat{\boldsymbol{\beta}} = 2(\mathbf{X}'\mathbf{Y} - \mathbf{X}'\mathbf{X}\hat{\boldsymbol{\beta}})[(\mathbf{Y} - \mathbf{X}\hat{\boldsymbol{\beta}})'(\mathbf{Y} - \mathbf{X}\hat{\boldsymbol{\beta}})]^{-1} \tag{2.6}$$

which can be written as

$$(\boldsymbol{\lambda}_0 + \mathbf{A}\hat{\boldsymbol{\beta}})(\mathbf{Y} - \mathbf{X}\hat{\boldsymbol{\beta}})'(\mathbf{Y} - \mathbf{X}\hat{\boldsymbol{\beta}}) = (\mathbf{X}'\mathbf{Y} - \mathbf{X}'\mathbf{X}\hat{\boldsymbol{\beta}}) \tag{2.7}$$

If $\boldsymbol{\lambda}_0 = 0$ and $\mathbf{A} = 0$, then (2.7) reduces to

$$\mathbf{X}'\mathbf{Y} = \mathbf{X}'\mathbf{X}\hat{\boldsymbol{\beta}} \tag{2.8}$$

which is the normal equation for estimation of $\hat{\boldsymbol{\beta}}$ giving the ordinary LSE. If $\boldsymbol{\lambda}_0 = 0$ but $\mathbf{A} \neq 0$, then (2.7) reduces to

$$\mathbf{A}\hat{\boldsymbol{\beta}}(\mathbf{Y} - \mathbf{X}\hat{\boldsymbol{\beta}})'(\mathbf{Y} - \mathbf{X}\hat{\boldsymbol{\beta}}) = \mathbf{X}'(\mathbf{Y} - \mathbf{X}\hat{\boldsymbol{\beta}}) \tag{2.9}$$

Suppose $\lambda(\boldsymbol{\beta})$ is chosen so that

$$\nabla \log \lambda(\boldsymbol{\beta}) = (2\boldsymbol{\lambda}_0 + 2\mathbf{A}\boldsymbol{\beta})[(\mathbf{Y} - \mathbf{X}\boldsymbol{\beta})'(\mathbf{Y} - \mathbf{X}\boldsymbol{\beta})]^{-1} \tag{2.10}$$

Here the weight function depends on the observation $\mathbf{Y}$, the design matrix $\mathbf{X}$ and the parameter $\boldsymbol{\beta}$. Then the equation (2.4) reduces to

$$\boldsymbol{\lambda}_0 + \mathbf{A}\boldsymbol{\beta} = \mathbf{X}'\mathbf{Y} - \mathbf{X}'\mathbf{X}\boldsymbol{\beta} \tag{2.11}$$

or equivalently

$$(\mathbf{X}'\mathbf{X} + \mathbf{A})\boldsymbol{\beta} = \mathbf{X}'\mathbf{Y} - \boldsymbol{\lambda}_0. \tag{2.12}$$

If the matrix $\mathbf{A}$ is chosen so that $\mathbf{X}'\mathbf{X} + \mathbf{A}$ is non-singular then the estimator given by relation (2.12) is the ridge estimator.

Let $\boldsymbol{\beta}^*$ denote the ordinary LSE. Note that

$$\mathbf{X}'\mathbf{Y} = \mathbf{X}'\mathbf{X}\boldsymbol{\beta}^* \tag{2.13}$$

In view of (2.4), it follows that

$$\nabla \log \lambda(\hat{\boldsymbol{\beta}}) = 2(\mathbf{X}'\mathbf{X}\boldsymbol{\beta}^* - \mathbf{X}'\mathbf{X}\hat{\boldsymbol{\beta}})[(\mathbf{Y} - \mathbf{X}\hat{\boldsymbol{\beta}})'(\mathbf{Y} - \mathbf{X}\hat{\boldsymbol{\beta}})]^{-1} \tag{2.14}$$

If $\mathbf{X}'\mathbf{X}$ is non-singular, then

$$\boldsymbol{\beta}^* - \hat{\boldsymbol{\beta}} = \tfrac{1}{2}[(\mathbf{Y} - \mathbf{X}\hat{\boldsymbol{\beta}}))'(\mathbf{Y} - \mathbf{X}\hat{\boldsymbol{\beta}})](\mathbf{X}'\mathbf{X})^{-1}\nabla \log \lambda(\hat{\boldsymbol{\beta}}) \tag{2.15}$$

and hence

$$\|\boldsymbol{\beta}^* - \hat{\boldsymbol{\beta}}\| \leqslant \tfrac{1}{2}[(\mathbf{Y} - \mathbf{X}\hat{\boldsymbol{\beta}})'(\mathbf{Y} - \mathbf{X}\hat{\boldsymbol{\beta}})]\|(\mathbf{X}'\mathbf{X})^{-1}\|\ \|\nabla \log \lambda(\hat{\boldsymbol{\beta}})\| \tag{2.16}$$

But,

$$\lambda(\hat{\boldsymbol{\beta}})[(\mathbf{Y} - \mathbf{X}\hat{\boldsymbol{\beta}})'(\mathbf{Y} - \mathbf{X}\hat{\boldsymbol{\beta}})] \leqslant \lambda(\boldsymbol{\beta}_0)[(\mathbf{Y} - \mathbf{X}\boldsymbol{\beta}_0)'(\mathbf{Y} - \mathbf{X}\boldsymbol{\beta}_0)]$$

since $\hat{\boldsymbol{\beta}}$ is a WLSE. Therefore,

$$\begin{aligned}\|\boldsymbol{\beta}^* - \hat{\boldsymbol{\beta}}\| &\leqslant \frac{1}{2}\frac{\lambda(\boldsymbol{\beta}_0)}{\lambda(\hat{\boldsymbol{\beta}})}[(\mathbf{Y} - \mathbf{X}\boldsymbol{\beta}_0)'(\mathbf{Y} - \mathbf{X}\boldsymbol{\beta}_0)]\|(\mathbf{X}'\mathbf{X})^{-1}\|\ \|\nabla \log \lambda(\hat{\boldsymbol{\beta}}\| \qquad (2.17)\\ &\leqslant c\|(\mathbf{X}'\mathbf{X})^{-1}\|[(\mathbf{Y} - \mathbf{X}\boldsymbol{\beta}_0)'(\mathbf{Y} - \mathbf{X}\boldsymbol{\beta}_0)]\\ &= c\|(\mathbf{X}'\mathbf{X})^{-1}\|\boldsymbol{\epsilon}'\boldsymbol{\epsilon}\end{aligned}$$

for some constant $c \geqslant 0$ provided

$$\inf_{\boldsymbol{\beta}} \lambda(\boldsymbol{\beta}) > 0$$

and

$$\sup_{\boldsymbol{\beta}} \|\nabla \log \lambda(\boldsymbol{\beta})\| < \infty.$$

We will study the properties of the estimator $\boldsymbol{\beta}^*$ in a general non-linear model in the next section. One can also use a matrix weight function $\Sigma(\boldsymbol{\beta})$ and study WLSE by minimizing

$$(\mathbf{Y} - \mathbf{X}\boldsymbol{\beta})'\Sigma(\boldsymbol{\beta})(\mathbf{Y} - \mathbf{X}\boldsymbol{\beta})$$

but we will not pursue this topic here.

2.3 NON-LINEAR REGRESSION

Consider the non-linear regression model

$$Y_i = g_i(\theta) + \epsilon_i,\ i \geqslant 1 \tag{3.0}$$

where ϵ_i, $i \geqslant 1$ are independent identically distributed random variables with mean 0 and variance σ^2 and $\theta \in \Theta \subset R$. $\hat{\theta}_n$ is said to be a weighted least squares estimator (WLSE) corresponding to the weight function $\lambda(\theta)$ if $\hat{\theta}_n$ minimizes

$$\nu_n(\theta) = \lambda(\theta) \sum_{i=1}^{n} (Y_i - g_i(\theta))^2 \tag{3.1}$$

over $\theta \in \bar{\Theta}$. Note that there exists a measurable WLSE if $\lambda(\theta)$ and $g_i(\theta)$ are continuous and $\bar{\Theta}$ is compact.

Suppose the following regularity conditions hold for $|\delta|$ near zero:

(R_1) $\sum_{i=1}^{n} [g_i(\theta_0) - g_i(\theta_0 + \delta)]^2 = Kn\delta^2 + o(1)$, and

(R_2) $\sum_{i=1}^{n} \epsilon_i[g_i(\theta_0) - g_i(\theta_0 + \delta)] = o_p(n)$.

(R_3) $\lambda(\cdot)$ is continuously differentiable in a neighbourhood of θ_0 with $\lambda'(\theta_0) = 0$ and $\inf_{\theta} \lambda(\theta) > 0$.

Under the conditions (R_1) and (R_2), it is easy to see that, for $\delta > 0$,

$$1/n\{\nu_n(\theta_0 \pm \delta) - \nu_n(\theta_0)\} \xrightarrow{p} (\lambda(\theta_0 \pm \delta) - \lambda(\theta_0))\sigma^2 + \lambda(\theta_0 \pm \delta)\delta^2 K \tag{3.2}$$

as $n \to \infty$. Note that, as $\delta \to 0$,

$$\lambda(\theta_0 \pm \delta) - \lambda(\theta_0)\sigma^2 + \lambda(\theta_0 \pm \delta)\delta^2 K \pm \delta\lambda'(\theta_0)\sigma^2 - \lambda(\theta_0)\delta^2 K \to 0$$

and

$$\pm\delta\lambda'(\theta_0)\sigma^2 + \lambda(\theta_0)\delta^2 K > 0$$

for all $\delta > 0$ since $\lambda'(\theta_0) = 0$ and $K > 0$.

Let $n > 0$. It follows that

$$\nu_n(\theta_0 \pm \delta) > \nu_n(\theta_0)$$

with probability greater than $1-\eta$ for n large for sufficiently small $\delta > 0$. In particular, it follows that $\nu_n(\theta)$ attains its minimum (say) at $\hat{\theta}_n \in (\theta_0 - \delta, \theta_0 + \delta)$ since $\nu_n(\theta)$ is continuous. Hence

$$P_{\theta_0}(|\hat{\theta}_n - \theta| < \delta) \geqslant 1 - \eta$$

for large n. In other words, $\hat{\theta}_n \xrightarrow{p} \theta_0$ as $n \to \infty$ and the following result holds.

Theorem 3.1. Under the conditions (R_1) to (R_3), there exists a locally weakly consistent estimator as $n \to \infty$.

Fix $\theta_0 \in \Theta$. Suppose the following regularity conditions hold:

(R_4) $\lambda(\theta)$ and $g_i(\theta)$ for $i \geqslant 1$ are twice continuously differentiable in a neighbourhood V_{θ_0} of θ_0 in Θ with $\inf_\theta \lambda(\theta) > 0$, $\lambda'(\theta_0) = 0$ and $\lambda''(\theta_0) \geqslant 0$.

(R_5) $\hat{\theta}_n$ is a weakly consistent solution of the least square equation

$$\nu_n'(\theta) = 0 \text{ or equivalently } \frac{d \log \lambda(\theta)}{d\theta} = \frac{2 \sum_{i=1}^{n} (Y_i - g_i(\theta)) g_i'(\theta)}{\sum_{i=1}^{n} (Y_i - g_i(\theta))^2} \tag{3.3}$$

(R_6) The sequence $\{\epsilon_i, i \geqslant 1\}$ and the sequence $\{g_i(\theta), i \geqslant 1\}$ are such that

$$(1/n) \sum_{i=1}^{n} \epsilon_i g_i''(\theta_0) \xrightarrow{p} 0 \text{ as } n \to \infty \tag{3.4}$$

(R_7) $\lim_{n\to\infty} n^{-1} \sum_{n=1}^{n} g_i'(\theta_0)^2 = K < \infty$ and $\lambda''(\theta_0)\sigma^2 + 2\lambda(\theta_0)K > 0$.

Theorem 3.2. Under the conditions (R_4) to (R_7),

(i) $\hat{\theta}_n \xrightarrow{p} \theta_0$ as $n \to \infty$

and

(ii) $\sqrt{n}(\hat{\theta}_n - \theta_0) \to N\left(0, \dfrac{4\lambda(\theta_0)^2 K \sigma^2}{(\lambda''(\theta_0)\sigma^2 + 2\lambda(\theta_0)K)^2}\right)$.

Proof. Note that

$$\nu_n(\theta) = \lambda(\theta) \sum_{i=1}^{n} (Y_i - g_i(\theta))^2 \tag{3.5}$$

and hence

$$\nu_n'(\theta) = \lambda'(\theta) \sum_{i=1}^{n} (Y_i - g_i(\theta))^2 - 2\lambda(\theta) \sum_{i=1}^{n} (Y_i - g_i(\theta)) g_i'(\theta), \tag{3.6}$$

$$\begin{aligned} \nu_n''(\theta) = {} & \lambda''(\theta) \sum_{i=1}^{n} (Y_i - g_i(\theta))^2 - 4\lambda'(\theta) \sum_{i=1}^{n} (Y_i - g_i(\theta)) g_i'(\theta) \\ & + 2\lambda(\theta) \sum_{i=1}^{n} g_i'(\theta)^2 - 2\lambda(\theta) \sum_{i=1}^{n} (Y_i - g_i(\theta)) g_i''(\theta). \end{aligned} \tag{3.7}$$

Since $\hat{\theta}_n \xrightarrow{p} \theta_0$ by (R_5), given $n > 0$, $\hat{\theta}_n \in V_{\theta_0}$ with probability greater than $1 - \eta$ for large n. In view of (R_4), $\nu_n''(\theta)$ is continuous in V_{θ_0} and

$$\nu_n'(\theta) = \nu_n'(\theta_0) + (\theta - \theta_0)[\nu_n''(\theta_0) + o(1)] \tag{3.8}$$

Since $\hat{\theta}_n \in V_{\theta_0}$ for large n, with probability greater than $1 - \eta$, it follows that

$$\nu_n'(\hat{\theta}_n) = \nu_n'(\theta_0) + (\hat{\theta}_n - \theta_0)[\nu_n''(\theta_0) + o_p(1)] \tag{3.9}$$

But, $\nu_n'(\hat{\theta}_n) = 0$ by (R_5). Hence

$$[\hat{\theta}_n - \theta_0][\nu_n''(\theta_0) + o_p(1)] = -\nu_n'(\theta_0) \tag{3.10}$$

In view of (R_4) to (R_7) and (3.7) it follows that

$$(1/n)\nu_n''(\theta_0) \xrightarrow{p} \lambda''(\theta_0)\sigma^2 + 2\lambda(\theta_0)K \text{ as } n \to \infty \tag{3.11}$$

since $\overline{\lim_n}\ (1/n) \sum_{i=1}^{n} \epsilon_i g_i'(\theta_0) < \infty$ in probability by WLLN applied to $\{\epsilon_i^2\}$ and the condition (R_7) using Cauchy-Schwartz inequality. Relations (3.10) and (3.11) show that

$$n^{1/2}(\hat{\theta}_n - \theta_0)[\{\lambda''(\theta_0)\sigma^2 + 2\lambda(\theta_0)K\} + o_p(1)] = -n^{-1/2}\nu_n(\theta_0) \tag{3.12}$$

But,

$$\nu_n'(\theta_0) = -2\lambda(\theta_0) \sum_{i=1}^{n} \epsilon_i g_i'(\theta_0) \tag{3.13}$$

Applying a central limit theorem due to Eicker (1963), it follows that

$$(1/\sqrt{n}) \sum_{i=1}^{n} \epsilon_i g_i'(\theta_0) \xrightarrow{\alpha} N(0, K\sigma^2)$$

provided,

$$\max_{1 \leq k \leq n} \frac{g_k'(\theta_0)^2}{\sum_{i=1}^{n} g_i'(\theta_0)^2} \to 0 \text{ as } n \to \infty$$

which in turn follows from Lemma 3 in Wu (1981), since

$$\lim_{n\to\infty} (1/n) \sum_{k=1}^{n} g_k'(\theta_0)^2 < \infty$$

by (R_7). Hence

$$(1/\sqrt{n})\nu_n'(\theta_0) \xrightarrow{\alpha} N(0, 4\lambda(\theta_0)^2\sigma^2 K) \tag{3.14}$$

Relations (3.12) and (3.14) prove that

$$n^{1/2}(\hat{\theta}_n - \theta_0) \xrightarrow{\alpha} N\left(0, \frac{4\lambda(\theta_0)^2 K\sigma^2}{(\lambda''(\theta_0)\sigma^2 + 2\lambda(\theta_0)K)^2}\right). \tag{3.15}$$

This completes the proof of the theorem.

Example: Consider the non-linear regression model

$$Y_i = g_i(\theta) + \epsilon_i,\ i \geq 1$$

where ϵ_i, $i \geqslant 1$ are i.i.d. random variables with mean 0 and variance σ_i^2. Suppose θ has a prior density $N(\theta_0, 1)$. Let

$$\lambda(\theta) = \exp\left(\tfrac{1}{2}(\theta - \theta_0)^2\right)$$

and $\widehat{\theta}_n$ denote the WLSE obtained by minimising

$$\nu_n(\theta) = \lambda(\theta) \sum_{i=1}^{n} [Y_i - g_i(\theta)]^2.$$

It is easy to see that

$$\inf_{\theta} \lambda(\theta) = 1,\ \lambda'(\theta_0) = 0,\ \lambda''(\theta_0) \geqslant 0$$

and hence the conditions on the weight function $\lambda(\theta)$ stated in (R_3) and (R_4) hold. Further if the functions $g_i(\cdot)$ are twice differentiable such that (R_1), (R_6) and (R_7) hold with $K > 0$ then, for some constant $\gamma > 0$,

$$\widehat{\theta}_n \xrightarrow{p} \theta_0 \text{ as } n \to \infty$$

and

$$n^{1/2}(\widehat{\theta}_n - \theta_0) \xrightarrow{\alpha} N(0, \gamma)$$

Observe that (R_2) is a consequence of (R_1) and the fact that $\{\epsilon_i\}$ are i.i.d. with mean zero and finite variance.

Remarks 3.1. Let $\widetilde{\theta}$ be the ordinary LSE. It is known that, and it follows from (3.15) by choosing $\lambda(\theta) \equiv 1$, that

$$n^{1/2}(\widetilde{\theta}_n - \theta_0) \xrightarrow{\alpha} N\left(0, \frac{\sigma^2}{K}\right)$$

under suitable conditions on $\{g_i(\theta), i \geqslant 1\}$ and $\{\epsilon_i, i \geqslant 1\}$. Observe that

$$\left\{\frac{4\lambda(\theta_0)^2\sigma^2K}{(\lambda''(\theta_0)\sigma^2 + 2\lambda(\theta_0)K)^2}\right\} < \frac{\sigma^2}{K} \tag{3.16}$$

if $\lambda''(\theta_0) > 0$ and hence the asymptotic variance of $\widehat{\theta}_n$ is ***strictly*** smaller than the asymptotic variance of $\widetilde{\theta}_n$ at θ_0. Note that if the weight function $\lambda(\cdot)$ is chosen so that $\lambda'(\theta_0) = 0$ and $\lambda''(\theta_0)$ is maximum, then the corresponding estimator will have the least asymptotic variance.

Remarks 3.2. Observe that, under the condition (R_4), $\lambda(\theta)$ attains its minimum at θ_0 if $\lambda''(\theta_0) > 0$. Suppose that $\lambda(\theta)$ has a global minimum at θ_0 over Θ. Now,

$$\begin{aligned}
\lambda(\theta_0) \sum_{i=1}^{n} \epsilon_i^2 &= \lambda(\theta_0) \sum_{i=1}^{n} (Y_i - g_i)(\theta_0))^2 \\
&\geqslant \lambda(\widehat{\theta}_n) \sum_{i=1}^{n} (Y_i - g_i(\widehat{\theta}_n))^2 \quad \text{(since } \widehat{\theta}_n \text{ is a WLSE)} \\
&\geqslant \lambda(\theta_0) \sum_{i=1}^{n} (Y_i - g_i(\widehat{\theta}_n))^2
\end{aligned}$$

since $\lambda(\theta)$ has minimum at $\theta = \theta_0$.
Hence

$$\sum_{i=1}^{n} \epsilon_i^2 \geqslant \sum_{i=1}^{n} (Y_i - g_i(\widehat{\theta}_n))^2$$

$$= \sum_{i=1}^{n} \epsilon_i^2 + n\psi_n(\widehat{\theta}_n, \theta_0) - 2nU_n(\widehat{\theta}_n)\psi_n(\widehat{\theta}_n, \theta_0)$$

where

$$\psi_n(\theta_1, \theta_2) = n^{-1} \sum_{i=1}^{n} (g_i(\theta_1) - g_i(\theta_2))^2,$$

and

$$U_n(\theta) = n^{-1} \sum_{i=1}^{n} \epsilon_i \left\{\frac{g_i(\theta) - g_i(\theta_0)}{\psi_n(\theta, \theta_0)}\right\}.$$

Note that, for any $\epsilon > 0$,

$$P_{\theta_0}[|\widehat{\theta}_n - \theta_0| > \epsilon] \leqslant P_{\theta_0}\{\sup_{|\theta-\theta_0|>\epsilon} U_n(\theta) \geqslant 1/2\}.$$

The following result is a consequence of arguments in Ivanov (1976) or Prakasa Rao (1984). We omit the proof.

Theorem 3.3. Suppose the following conditions hold.

(R_8) $\lambda(\theta)$ attains its minimum at θ_0.

(R_9) There exist constants k_1 and k_2 such that

$$k_1(\theta_1 - \theta_2)^2 \leqslant \psi_n(\theta_1, \theta_2) \leqslant k_2(\theta_1 - \theta_2)^2.$$

(R_{10}) $E|\epsilon_1|^s < \infty$ for some $s \geqslant 2$.

Then there exists a constant $c_s > 0$ such that, for all $\rho > 0$,

$$P_{\theta_0}(n^{1/2}|\widehat{\theta}_n - \theta_0| > \rho) \leqslant c_s\rho^{-s}, \; n \geqslant 1 \tag{3.17}$$

In particular, if $s > 2$, then it follows that

$$\widehat{\theta}_n \xrightarrow{\text{a.s.}} \theta_0 \text{ as } n \to \infty \tag{3.18}$$

by Borel-Cantelli Lemma.

Since the asymptotic properties of LSE are of limited interest in practice in the scalar parameter case, we now extend the results obtained above to the vector parameter case.

Let us now consider again the non-linear regression model

$$Y_i = g_i(\theta) + \epsilon_i, \; i \geqslant 1 \tag{3.19}$$

where ϵ_i, $i \geqslant 1$ are independent random variables with $E(\epsilon_i) = 0$ and $\text{Var}(\epsilon_i) = \sigma_i^2 < \infty$. Suppose $\theta \in \Theta \subset R^k$. $\widehat{\theta}_n$ is said to be a weighted least squares estimator (WLSE) corresponding to the positive weight function $\lambda(\theta)$ if $\widehat{\theta}_n$ minimizes

$$\nu_n(\theta) = \lambda(\theta) \sum_{i=1}^{n} (Y_i - g_i(\theta))^2$$

over $\theta \in \bar{\Theta}$.

Suppose the following regularity conditions hold whenever $\mathbf{h}$ is such that $\|\mathbf{h}\|$ is small.

(R_1') There exists a positive constant K^* such that

$$\sum_{i=1}^{n} [g_i(\theta_0) - g_i(\theta_0 + \mathbf{h})]^2 = K^* n \|\mathbf{h}\|^2 + o(1);$$

(R_2') $\sum_{i=1}^{n} \epsilon_i [g_i(\theta_0) - g_i(\theta_0 + \mathbf{h})] = o_p(n); \frac{1}{n} \sum_{i=2}^{n} \epsilon_i^2 \xrightarrow{p} \sigma^2$ as $n \to \infty$;

(R_3') $\lambda(\cdot)$ is continuously totally differentiable in a neighbourhood of θ_0 with $\nabla\lambda(\theta_0) = 0$.

It is easy to check that

$$(1/n)[\nu_n(\theta_0 + \mathbf{h}) - \nu_n(\theta_0)] \xrightarrow{p} (\lambda(\theta_0 + \mathbf{h}) - \lambda(\theta_0))\sigma^2 + \lambda(\theta_0 + \mathbf{h})\|\mathbf{h}\|^2 K^*$$

as $n \to \infty$. Furthermore,

$$[\lambda(\theta_0 + \mathbf{h}) - \lambda(\theta_0)]\sigma^2 + \lambda(\theta_0 + \mathbf{h})\|\mathbf{h}\|^2 K^* - \mathbf{h}'\nabla\lambda(\theta_0)\sigma^2 - \lambda(\theta_0)\|\mathbf{h}\|^2 K^* \to 0$$

as $\|\mathbf{h}\| \to 0$ by (R_3'). Since $\nabla\lambda(\theta_0) = 0$, it follows that

$$[\lambda(\theta_0 + \mathbf{h}) - \lambda(\theta_0)]\sigma^2 + \lambda(\theta_0 + \mathbf{h})\|\mathbf{h}\|^2 K^* - \lambda(\theta_0)\|\mathbf{h}\|^2 K^* \to 0$$

as $\|\mathbf{h}\| \to 0$. Let $\eta > 0$. Since $\lambda(\theta_0) > 0$, it follows that

$$\nu_n(\theta_0 + \mathbf{h}) > \nu_n(\theta_0)$$

with probability greater than $1 - \eta$ for n large for $\mathbf{h}$ such that $\|\mathbf{h}\| = \delta$ for sufficiently small $\delta > 0$. Since $\nu_n(\theta)$ is continuous, it attains its minimum over the sphere $B(\theta_0, \delta) = \{\theta : \|\theta - \theta_0\| \leqslant \delta\}$. But the minimum is not attained on the boundary $\partial B(\theta_0, \delta)$ since $\nu_n(\theta_0 + \mathbf{h}) > \nu_n'(\theta_0)$ for $\|\mathbf{h}\| = \delta$. Hence

$$P_{\theta_0}(\|\hat{\theta}_n - \theta_0\| < \delta) > 1 - \eta$$

for large n. In other words $\hat{\theta}_n \xrightarrow{p} \theta_0$ as $n \to \infty$ and we have the following result.

Theorem 3.4. Under the conditions (R_1') to (R_3'), there exists a locally weakly consistent estimator.

Fix $\theta_0 \in \Theta$. Suppose the following regularity conditions hold.

(R_4') $\lambda(\theta)$ and $g_i(\theta)$, $i \geqslant 1$ are twice continuously totally differentiable in a neighbourhood V_{θ_0} of θ_0 in Θ with $\nabla\lambda(\theta_0) = \mathbf{0}$ and $\nabla^2\lambda(\theta_0)$ positive definite.

(R_5') $\hat{\theta}_n \xrightarrow{p} \theta_0$ as $n \to \infty$ and $\hat{\theta}_n$ is a solution of the equation

$$\nabla\nu_n(\theta) = 0$$

or equivalently

$$\nabla \log \lambda(\theta) = \frac{2 \sum_{i=1}^{n} (Y_i - g_i(\theta))\nabla g_i(\theta)}{\sum_{i=1}^{n} (Y_i - g_i(\theta))^2}.$$

(R_6') The sequence $\{\epsilon_i, i \geqslant 1\}$ and the sequence of $\{g_i(\theta), i \geqslant 1\}$ are such that $E(\epsilon_i) = 0$, Var $(\epsilon_i) = \sigma_i^2 < \infty$ and $\inf_i \sigma_i^2 > 0$. Further,

(i) $(1/n) \sum_{i=1}^{n} \epsilon_i \nabla^2 g_i(\theta_0) \xrightarrow{p} \mathbf{0}$ as $n \to \infty$.

(ii) $(1/n) \sum_{i=1}^{n} \sigma_i^2 \to \sigma^2 > 0$ as $n \to \infty$.

(R_7') $\lim_{n\to\infty} (1/n) \sum_{i=1}^{n} \nabla g_i(\theta_0) \nabla g_i(\theta_0)' = \mathbf{K}$ and

$$\lim_{n\to\infty} (1/n) \sum_{i=1}^{n} \nabla g_i(\theta_0) \nabla g_i(\theta_0)' \sigma_i^2 = \mathbf{K}^*$$

where K and K^* are non-negative definite matrices such that

$$M = \nabla^2 \lambda(\theta_0) \sigma^2 + 2\lambda(\theta_0)\mathbf{K}$$

is positive definite.

Theorem 3.5. Under the conditions (R_4') to (R_7'),

(i) $\widehat{\theta}_h \xrightarrow{p} \theta_0$ as $n \to \infty$,

and

(ii) $\sqrt{n}(\widehat{\theta}_n - \theta_0) \xrightarrow{\alpha} N_k(\mathbf{0}, \Sigma)$,

where $\Sigma = (\mathbf{M}^{-1}\mathbf{K}^*\mathbf{M}^{-1})4\lambda^2(\theta_0)$ and

$\mathbf{M} = \nabla^2\lambda(\theta_0)\sigma^2 + 2\lambda(\theta_0)\mathbf{K}$.

Proof. We sketch the proof as it is similar to the proof of Theorem 3.2 in the scalar case. Note that, for large n, with probability tending to one,

$$\nabla \nu_n(\widehat{\theta}_n) = \nabla \nu_n(\theta_0) + [\nabla^2 \nu_n(\theta_0) + o_p(1)](\widehat{\theta}_n - \theta_0).$$

Since, $\nabla \nu_n(\widehat{\theta}_n) = 0$, it follows that

$$[\nabla^2 \nu_n(\theta_0) + o_p(1)](\widehat{\theta}_n - \theta_0) = -\nabla \nu_n(\theta_0).$$

It is easy to see that

$$(1/n)\nabla^2 \nu_n(\theta_0) \xrightarrow{p} \nabla^2\lambda(\theta_0)\sigma^2 + 2\lambda(\theta_0)\mathbf{K} = \mathbf{M}$$

where $\mathbf{K}$ is as defined in (R_7'). This follows from (R_6') and (R_7'). On the other hand,

$$n^{1/2}\nabla \nu_n(\theta_0) \xrightarrow{\alpha} N_k(0, 4\lambda(\theta_0)^2\mathbf{K}^*)$$

by Eicker (1963) in view of (R_6') and (R_7') since

$$\nabla \nu_n(\theta_0) = -2\lambda(\theta_0) \sum_{i=1}^{n} \epsilon_i \nabla g_i(\theta_0).$$

Hence, applying standard arguments, it follows that

$$n^{1/2}(\widehat{\theta}_n - \theta_0) \xrightarrow{\alpha} N_k(0, M^{-1}(4\lambda(\theta_0)^2\mathbf{K}^*)\mathbf{M}^{-1}).$$

Corollary 3.1. Consider the linear model

$$\mathbf{Y} = \mathbf{X}\boldsymbol{\beta} + \boldsymbol{\epsilon}$$

defined in (2.0) where $\mathbf{Y}$ is $k \times 1$ vector of observations, $\mathbf{X}$ is a design matrix of order $k \times m$ and $\boldsymbol{\beta}$ is a $m \times 1$ vector of unknown parameters. Suppose $\boldsymbol{\epsilon}$ is a $k \times 1$ random vector with $E(\epsilon_i) = 0,\ 1 \leqslant i \leqslant M$ where ϵ_i is the i-th component. Let $\mathbf{X}_i$ be the i-th row-vector of the matrix $\mathbf{X}$. Assume $\{\epsilon_i\}$ is a sequence of independent random variables such that $\text{Var}(\epsilon_i) = \sigma_i^2 < \infty$, $\inf_i \sigma_i^2 > 0$ and

$$\frac{1}{n} \sum_{i=1}^{n} \epsilon_i^2 \xrightarrow{p} \sigma^2 > 0 \tag{3.20}$$

as $n \to \infty$. Let $\widehat{\boldsymbol{\beta}}_k$ be the ordinary LSE based on $\mathbf{Y}$.

If

$$\frac{1}{k} \sum_{i=1}^{k} \mathbf{X}_i' \mathbf{X}_i \to \mathbf{K} \text{ as } k \to \infty \tag{3.21}$$

where $\mathbf{K}$ is a positive definite matrix, then

$$\widehat{\boldsymbol{\beta}}_k \xrightarrow{p} \boldsymbol{\beta} \quad \text{as } k \to \infty.$$

Further, if in addition to (3.21),

$$\frac{1}{k} \sum_{i=1}^{k} \mathbf{X}_i' \mathbf{X}_i \sigma_i^2 \to \mathbf{K}^* \text{ as } k \to \infty \tag{3.22}$$

where $\mathbf{K}^*$ is a positive definite matrix, then

$$k^{1/2}(\widehat{\boldsymbol{\beta}}_k - \boldsymbol{\beta}) \xrightarrow{\alpha} N_m(0, \Sigma) \quad \text{as } k \to \infty$$

where $\Sigma = \mathbf{K}^{-1}\mathbf{K}^*\mathbf{K}^{-1}$.

Remark 3.3. Observe that, for a least square estimator $\widetilde{\theta}_n$,

$$n^{1/2}(\widetilde{\theta}_n - \theta_0) \xrightarrow{\alpha} N_k(0, \mathbf{K}^{-1}\mathbf{K}^*\mathbf{K}^{-1}).$$

Under suitable conditions on $\{g_i(\theta), i > 1\}$ and $\{\epsilon_i, i \geqslant 1\}$, it can be checked that

$$\mathbf{K}^{-1}\mathbf{K}^*\mathbf{K}^{-1} - \mathbf{M}^{-1}\mathbf{K}^*\mathbf{M}^{-1}\ 4\lambda(\theta_0)^2$$

is positive definite and hence the asymptotic covariance matrix of $\widehat{\theta}_n$ is strictly smaller than that of θ_n as long as $\nabla^2\lambda(\theta_0)$ is positive definite.

Remark 3.4. One of the basic assumptions in asymptotic theory of non-linear regression is that

$$\lim_{n\to\infty} (1/n) \sum_{i=1}^{n} g_i'^2(\theta_0) > 0$$

in the scalar parameter case and

$$\lim_{n\to\infty} (1/n) \sum_{i=1}^{n} \nabla g_i(\theta_0) \nabla g_i(\theta_0)'$$

is positive definite in the vector parameter case. These are analogues of the condition that $\mathbf{X}'\mathbf{X}$ is non-singular in the standard linear model $\mathbf{Y} = \mathbf{X}\boldsymbol{\beta} + \boldsymbol{\epsilon}$.

In the light of the approach discussed by us, we can now obtain a type of ridge-estimator for non-linear regression models when the conditions stated above do not hold. Consider the non-linear regression model

$$Y_i = g_i(\theta) + \epsilon_i, \ i \geqslant 1$$

and θ_0 be an initial value for the scalar parameter θ. Consider a linearized version of the above model in a neighbourhood of θ_0.

$$Y_i = g_i(\theta_0) + (\theta - \theta_0)g_i'(\theta_0) + \epsilon_i, \ i \geqslant 1.$$

Let $\lambda(\theta)$ be a positive weight function. Suppose $\lambda(\cdot)$ is differentiable.

Let $\widehat{\theta}$ be a weighted approximate **LSE (WLSE)** in the sense that it minimizes the approximate weighted squares

$$\lambda(\theta) \sum_{i=1}^{n} [Y_i - g_i(\theta_0) - (\widehat{\theta} - \theta_0)g_i'(\theta_0)]^2.$$

Hence,

$$\lambda'(\widehat{\theta}) \sum_{i=1}^{n} [Y_i - g_i(\theta_0) - (\widehat{\theta} - \theta_0)g_i'(\theta_0)]^2$$

$$= 2\lambda(\widehat{\theta}) \sum_{i=1}^{n} [Y_i - g_i(\theta_0) - (\widehat{\theta} - \theta_0)g_i'(\theta_0)]g_i(\theta_0)$$

or in other words

$$\frac{d \log \lambda(\widehat{\theta})}{d\theta} = \frac{2 \sum_{i=1}^{n} [Y_i - g_i(\theta_0) - (\widehat{\theta} - \theta_0)g_i'(\theta_0)]g_i'(\theta_0)}{\sum_{i=1}^{n} [Y_i - g_i(\theta_0) - (\widehat{\theta} - \theta_0)g_i'(\theta_0)]^2}.$$

Choose $\lambda(\theta)$ such that

$$\frac{d \log \lambda(\theta)}{d\theta} = \frac{2(na + nb(\theta - \theta_0))}{\sum_{i=1}^{n} [Y_i - g_i(\theta_0) - (\theta - \theta_0)g_i'(\theta_0)]^2}.$$

Then, it follows that

$$na + nb(\widehat{\theta} - \theta_0) = \sum_{i=1}^{n} [Y_i - g_i(\theta_0) - (\widehat{\theta} - \theta_0)g_i'(\theta_0)]g_i'(\theta_0)$$

and hence,

$$\widehat{\theta} - \theta_0 = \frac{(1/n) \sum_i (Y_i - g_i(\theta_0))g_i'(\theta_0) - a}{(1/n) \sum_i g_i'^2(\theta_0) + b}.$$

In the vector parameter case, similar analysis leads to estimators of the form

$$\widehat{\theta} - \theta_0 = \left\{n^{-1} \sum_{i=1}^{n} (Y_i - g_i(\theta_0))\nabla g_i(\theta_0) - \boldsymbol{\alpha}\right\}\{n^{-1} \sum_i \nabla g_i(\theta_0)\nabla g_i(\theta_0)' + \mathbf{B}\}^{-1}$$

where $\mathbf{B}$ is chosen to be a positive definite matrix and $\boldsymbol{\alpha}$ is a vector. This is a type of ridge estimator. The above relation will give an iterative procedure for computing an estimator of θ. The asymptotic properties of $\hat{\theta}$ remain to be investigated. The advantage of this procedure is that it can be used even when

$$(1/n) \sum_{i=1}^{n} \nabla g_i(\theta_0) \nabla g_i(\theta_0)'$$

converges to a singular matrix, since **B** is positive definite.

2.4. AUTOREGRESSIVE TIME SERIES

Let us first consider the first order autoregressive time series defined by

$$Y_n = \rho Y_{n-1} + \epsilon_n,\ n \geqslant 1,\ Y_0 = 0 \tag{4.0}$$

where $\{\epsilon_i, i \geqslant 1\}$ are i.i.d. random variables with mean 0. The problem here is estimation of ρ based on $Y_1, \ldots, Y_n$.

Let us consider

$$\nu_n(\rho) = \lambda(\rho) \sum_{i=1}^{n} (Y_i - \rho Y_{i-1})^2 \tag{4.1}$$

where $\lambda(\rho)$ is a given weight function and choose $\hat{\rho}_n$ minimizing $\nu_n(\rho)$ as an estimator for ρ. It is easy to see that $\hat{\rho}_n$ is a solution of the equation

$$\frac{d \log \lambda(\rho)}{d\rho} = \frac{2 \sum_{i=1}^{n} Y_{i-1}(Y_i - \rho Y_{i-1})}{\sum_{i=1}^{n} (Y_i - \rho Y_{i-1})^2} \tag{4.2}$$

Suppose we choose $\lambda(\rho)$ such that

$$\frac{d \log \lambda(\rho)}{d\rho} = \frac{n(2a + 2b\rho)}{\sum_{i=1}^{n} (Y_i - \rho Y_{i-1})^2} \tag{4.3}$$

Then (4.2) reduces to

$$n(a + b\rho) = \sum_{i=1}^{n} (Y_{i-1}(Y_i - \rho Y_i))$$

and hence,

$$\hat{\rho}_n = \frac{\sum_{i=1}^{n} Y_i Y_{i-1} - an}{\sum_{i=1}^{n} Y_{i-1}^2 - bn} = \frac{1/n \sum_{i=1}^{n} Y_i Y_{i-1} - a}{1/n \sum_{i=1}^{n} Y_{i-1}^2 - b} \tag{4.4}$$

This is an analogue of ridge estimator in the linear model.

Let us consider p-th order autoregressive model defined by

$$Y_n = \rho_0 + \rho_1 Y_{n-1} + \ldots + \rho_p Y_{n-p} + \epsilon_n,\ n \geqslant p + 1,\ Y_0 \equiv 0 \tag{4.5}$$

where $\{\epsilon_i, i \geqslant 1\}$ are i.i.d. random variables with mean 0. The problem again in the estimation of $\rho' \equiv (\rho_0, \rho_1, \ldots, \rho_p)$ based on $Y_1, \ldots, Y_p, Y_{p+1}, \ldots, Y_n$.

Let us consider

$$\nu_n(\rho) = \lambda(\rho) \sum_{i=p+1}^{n} (Y_i - \rho_0 Y_{i-1} - \ldots - \rho_p Y_{i-p})^2$$

$$= \lambda(\rho) \sum_{i=p+1}^{n} (Y_i - \mathbf{Y}_i'\boldsymbol{\rho})^2 \tag{4.6}$$

where $\mathbf{Y}_n' = (Y_{n-1}, \ldots, Y_{n-p})$. Differentiating (4.6) with respect to ρ we have

$$\nabla\lambda(\rho) \sum_{i=p+1}^{n} (Y_i - \mathbf{Y}_i'\boldsymbol{\rho})^2 = 2\lambda(\rho) \sum_{i=p+1}^{n} \mathbf{Y}_i(Y_i - \mathbf{Y}_i'\boldsymbol{\rho})$$

or in other words

$$\nabla \log \lambda(\rho) = \frac{2 \sum_{i=p+1}^{n} \mathbf{Y}_i(Y_i - \mathbf{Y}_i'\boldsymbol{\rho})}{\sum_{i=p+1}^{n} (Y_i - \mathbf{Y}_i'\boldsymbol{\rho})^2}.$$

In particular, if

$$\nabla \log \lambda(\rho) = \frac{(2\lambda + 2A\boldsymbol{\rho})(n-p)}{\sum_{i=p+1}^{n} (Y_i - \mathbf{Y}_i'\boldsymbol{\rho})^2}$$

then

$$(n-p)(2\boldsymbol{\lambda} + 2A\hat{\boldsymbol{\rho}}) = 2 \sum_{i=p+1}^{n} \mathbf{Y}_i(Y_i - \mathbf{Y}_i'\hat{\rho})$$

and hence,

$$\hat{\boldsymbol{\rho}} = \left\{\sum_{i=p+1}^{n} \mathbf{Y}_i Y_i + (n-p)\boldsymbol{\lambda}\right\}\left\{\sum_{i=p+1}^{n} \mathbf{Y}_i\mathbf{Y}_i' + (n-p)\mathbf{A}\right\}^{-1}$$

$$= \left\{\frac{1}{n-p} \sum_{i=p+1}^{n} \mathbf{Y}_i Y_i' + \boldsymbol{\lambda}\right\}\left\{\frac{1}{n-p} \sum_{i=p+1}^{n} \mathbf{Y}_i\mathbf{Y}_i' + \mathbf{A}\right\}^{-1}.$$

Let,

$$\mathbf{J}_n = \frac{1}{n-p} \sum_{i=p+1}^{n} \mathbf{Y}_i\mathbf{Y}_i'.$$

If $\mathbf{J}_n \xrightarrow{p} \mathbf{J}$ where $\mathbf{J}$ is possibly singular, then $\hat{\boldsymbol{\rho}}$ can be taken as a type of ridge estimator of $\boldsymbol{\rho}$. Here $\mathbf{A}$ should be chosen so that $\mathbf{J} + \mathbf{A}$ is a non-singular matrix. The problem of study of asymptotic properties of $\hat{\boldsymbol{\rho}}$ remains to be investigated.

Acknowledgement. The author thanks the referee for his detailed comments and suggestions.

REFERENCES

1. Basawa, I. V. and Prakasa Rao, B.L.S. (1980): *Statistical Inference for Stochastic Processes,* Academic Press, London.
2. Eicker, F. (1963): Central limit theorems for families of sequences of random variables. *Ann. Math Statist,* **34,** 439–446.
3. Ivandv, A. V. (1976): An asymptotic expansion for the distribution of the least squares estimator of the non-linear regression parameter, *Theory Probab. Appl.,* **21,** 557–570.

4. Jennrich, R. I. (1969): Non-linear least squares estimation. *Ann. Math. Statist.*, **40**, 633–643.

5. Prakasa Rao, B. L. S. (1984): The rate of convergence of the least squares estimator in a non-linear regression model with dependent errors. *J. Multivariate Anal.*, **14**, 315–322.

6. ——— (1985): *Asymptotic theory of estimation in non-linear regression*, (preprint) Indian Statistical Institute, New Delhi.

7. Rao, C. R. (1973): *Linear Statistical Inference and its Applications*, Wiley, New York.

8. W. C. F. (1981): Asymptotic theory of non-linear least squares estimation. *Ann. Statistics*, **9**, 501–513.

3

Matrix Beta Distribution with Applications to Linear Models, Testing, Skewness and Kurtosis

C. G. Khatri

Gujarat University, Ahmedabad

3.1 INTRODUCTION

We shall denote $\mathcal{R}^{p\times n}$ as a set of $p\times n$ matrices defined on the real space $\mathcal{R}$ and $\mathcal{R}^p = R^{p\times l}$. Let U be a $p\times n$ matrix such that $UU' = \mathbf{I}_p$ and the set of such matrices are denoted by $o(p, n)$, a Stiefel's manifold. $X \overset{d}{=} Y$ implies that X and Y are identically distributed.

We shall say that $X(\epsilon\mathcal{R}^{p\times n})$ has a right spherical non-singular distribution if $X \overset{d}{=} XA$ for all orthogonal matrices $A\in o(n)$ and $P(XX'>0) = 1$. Then, according to Dawid (1977), $X \overset{d}{=} TU$ where T and U are independently distributed, $S = TT' = XX'$ and U is uniformly distributed over $O(p, n)$. Hence, it is easy to see that

$$R = X'(XX')^{-1}X \overset{d}{=} U'U \tag{1.1}$$

and according to Mitra (1970) or Khatri (1970), $R = U'U$ is said to have a matrix beta distribution, denoted by $\beta_n(\frac{1}{2}p,\ \frac{1}{2}(n-p))$ and note that its density function does not exist. The moments of $R = U'U$ can be obtained using the results of James (1964). Let $U_0' = (U', U_1')$ be an orthogonal matrix. Then $R = U'U = U_0'BU_0$ with $B = \begin{pmatrix} I_p & 0 \\ 0 & 0\end{pmatrix}$ being $n\times n$ matrix, and using James (1964), we get

$$\int_{O(n)} C_k(AU'U)[dU_0] = \int_{O(n)} C_k(AU_0'BU')[dU_0] = \frac{C_k(A)C_k(B)}{C_k(I_n)}$$
$$= (\tfrac{1}{2}p)_k C_\kappa(A)/(\tfrac{1}{2}n)_k \tag{1.2}$$

where κ is a partition $\{k_1, k_2, \dots, k_p\}$ of k into p parts, $k_1\geqslant k_2\geqslant\dots\geqslant k_p\geqslant 0$ and $\sum_{i=1}^{p} k_i = k$, $[dU_0]$ = unit invariant Haar measure and $C_k(\cdot)$ denotes the zonal polynomial corresponding to κ [for details, see James (1964)].

Notice that $U \overset{d}{=} B_0UA'$ for any orthogonal matrices B_0 $(\epsilon O(p))$ and $A \in O(n)$. If $A \in O(q, n)$. then $U_1 = UA'$ for $n > p + q$ is distributed as

$$\Gamma_p(n/2)\left\{\pi^{(1/2)pq}\Gamma_p\left(\frac{n-q}{2}\right)\right\}^{-1}|I_p - U_1U_1'|^{(1/2)(n-p-q-1)} \tag{1.3}$$

for all $U_1 \in \mathcal{R}^{p\times q}$ with $I_p - U_1U_1' > 0$, [see Khatri (1970)], and $ARA' \overset{d}{=} U_1'U_1 \sim \beta_q\left(\frac{1}{2}p, \frac{n-p}{2}\right)$ and if $q \leqslant p$, then its density is given by

$$\{\beta_q(\tfrac{1}{2}p, \tfrac{1}{2}(n-p))\}^{-1}|R_1|^{(1/2)(p-q-1)}|I_q - R_1|^{(1/2)(n-p-q-1)} \tag{1.4}$$

for all $0 < R_1 = ARA' < I_q$, where

$$\beta_q(a, b) = \Gamma_q(a)\Gamma_q(b)/\Gamma_q(a+b) \text{ and } \Gamma_q(a) = \Pi^{q(q-1)/4}\prod_{j=1}^{q}\Gamma\left(a - \frac{j-1}{2}\right) \tag{1.5}$$

In Section 3.2, we study the linear model.

$$\mathbf{Y} = X_1'\boldsymbol{\beta}_1 + X_2'\boldsymbol{\beta}_2 + \boldsymbol{\epsilon} \tag{1.6}$$

where $\boldsymbol{\epsilon}$ and X_2 are independently distributed, X_2 has a right spherical distribution, $E\boldsymbol{\epsilon} = \mathbf{0}$, $V(\boldsymbol{\epsilon}) = \Sigma$, X_1 and $\mathbf{Y}$ are known matrices, and $\boldsymbol{\beta}_1$ and $\boldsymbol{\beta}_2$ are unknown parameters. Let $I_n - X_1'(X_1X_1')^{-1}X_1 = X_3'X_3$ with $X_3X_3' = I_{n-r}$, $X_{(2)} = X_2X_3'$, $\mathbf{Y}_1 = X_3\mathbf{Y}$ and $\Sigma_1 = X_3\Sigma X_3'$. Then, the two least squares estimates of $\boldsymbol{\beta}_2$ are

$$\widetilde{\boldsymbol{\beta}}_2 = (X_{(2)}X_{(2)}')^{-1}X_{(2)}\mathbf{Y}_1 \text{ and } \widehat{\boldsymbol{\beta}}_2 = \{x_{(2)}\Sigma_1^{-1}X_{(2)}')^{-1}X_{(2)}\Sigma_1^{-1}\mathbf{Y} \tag{1.7}$$

with

$$(V\widetilde{\boldsymbol{\beta}}_2 \mid X_2) = [X_{(2)}X_{(2)}']^{-1}X_{(2)}\Sigma_1X_{(2)}'[X_{(2)}X_{(2)}']^{-1}$$

and

$$V(\widehat{\boldsymbol{\beta}}_2 \mid X_2) = [X_{(2)}\Sigma_1^{-1}X_{(2)}']^{-1} \tag{1.8}$$

Define the conditional efficiency factor by

$$e_k = tr_k[\{V(\widehat{\boldsymbol{\beta}}_2 \mid X_2)\}^{-1}V(\widetilde{\boldsymbol{\beta}}_2 \mid X_2)]\Big/\binom{p}{k}$$

$$= tr_k(\Sigma_1^{-1}R\Sigma_1R)\Big/\binom{p}{k} \tag{1.9}$$

where $R = X_{(2)}'[X_{(2)}X_{(2)}']^{-1}X_{(2)} \overset{d}{=} U'U \sim \beta_{n-r}(\frac{1}{2}p, \frac{1}{2}(n-r-p))$ or U is uniformly distributed over $O(p, n-r)$ and $tr_k(\cdot)$ denotes the sum of the principal minors of order $k\times k$ from $(\cdot)$. Then, Section 3.2 gives the average efficiency of e_k or $E(e_k)$. When $X_1 = 0$ and $k = p$, the result was given by Iwasaki (1985).

In Section 3.3, we consider the linear model

$$X = \xi C' + \epsilon \tag{1.10}$$

where $C \in \mathcal{R}^{n\times c}$ is a known matrix of rank C, ξ is unknown and ϵ has a right spherical distribution and $P(\epsilon\epsilon' > 0) = 1$. Some test matrices are given so that its null distribution is beta for testing the hypothesis $H_0(\xi B' = 0)$ against $H(\xi B' \neq 0)$.

In the last Section 3.4, we consider the linear model given by (1.10) and define Mardia's measures of skewness and kurtosis (1970, 1974) and obtain their exact moments. Note that the least squares estimate of $\xi C'$ is $\widehat{X} = \widehat{\xi}C' = XC(C'C)^{-1}C' = (\widehat{\mathbf{x}}_1, \ldots, \widehat{\mathbf{x}}_n)$ and let $X = (\mathbf{x}_1, \ldots, \mathbf{x}_n)$. Define

$$S = \sum_{i=1}^{n} (\mathbf{x}_i - \widehat{\mathbf{x}}_i)(\mathbf{x}_i - \widehat{\mathbf{x}}_i)' \Big/ n, \; R_{ij} = (\mathbf{x}_i - \widehat{\mathbf{x}}_i)' S^{-1} (\mathbf{x}_j - \widehat{\mathbf{x}}_j)$$

for $i, j = 1, 2, \ldots, n$. The sample distances R_{ii} $(i = 1, 2, \ldots, n)$ can be utilized in the study of clustering the points, and these are used in defining skewness and kurtosis as

$$b_{1p} = \sum_{i,j=1}^{n} R_{ij}^3 \Big/ n^2 \quad \text{and} \quad b_{2p} = \sum_{i=1}^{n} R_{ii}^2 \Big/ n \tag{1.11}$$

[see Mardia (1970, 1974)]. Section 3.4 gives the exact expressions for Eb_{1p}, Eb_{2p} and Eb_{2p}^2. Other moments can be determined.

3.2. AVERAGE EFFICIENCY OF e_k

We refer to the definition of e_k given by (1.9), and assume that $k \leqslant \min(p, n - r)$. Let $n - r = m$. If $m - k < k$, then $Ee_k = Ee_{m-k}$. Hence, $k \leqslant (m/2, p)$. We develop the following:

Theorem 1. Let $a_k(A, B) = tr_k(ARBR)$ where A and B are $m \times m$ symmetric matrices and $R = U'U \sim \beta_m(\frac{1}{2}p, \frac{1}{2}(m-p))$.

Let $A = \Delta D_a \Delta'$, $B = \Delta B_0 \Delta'$ and Δ be an orthogonal matrix with $D_a = \text{diag}(a_1, a_2, \ldots, a_m)$. Let Σ_l^* denote the summation over the distinct possible combinations of $(k + l)$ elements $\{\alpha_1, \ldots, \alpha_{k-l}, \beta_1, \ldots, \beta_l, \gamma_1, \ldots, \gamma_l\}$ out of $1, 2, \ldots, m$ and the total number of terms are

$$\binom{m-2l}{k-l}\binom{m}{2l}\binom{2l}{l} \text{ with } \alpha_1 < \alpha_2 < \ldots < \alpha_{k-l}, \beta_1 < \beta_2 < \ldots < \beta_l$$

and $\gamma_1 < \gamma_2 < \ldots < \gamma_l$. Further, let $B_0(\boldsymbol{\alpha}, \boldsymbol{\beta}; \boldsymbol{\alpha}, \boldsymbol{\gamma})$ be a $k \times k$ matrix obtained by taking $(\boldsymbol{\alpha}, \boldsymbol{\beta})$ row vectors and $(\boldsymbol{\alpha}, \boldsymbol{\gamma})$ column vectors from B_0. Then

$$Ea_k(A, B) = \sum_{l=1}^{k} W_l \bar{A}_l(A, B) \binom{p}{k} \tag{2.1}$$

where,

$$W_l = \binom{p+2}{p-k+l+2}\binom{m-p}{l} \Big/ \binom{m+2}{k} \quad \text{for } l = 1, 2, \ldots, k \tag{2.2}$$

and

$$\bar{A}_l(A, b) = \frac{\Sigma_l^* \left(\prod_{t=1}^{k-l} a_{\alpha_t}\right)\left(\prod_{t=1}^{l} a_{r_t}\right) \Big/ |B_0(\boldsymbol{\alpha}, \boldsymbol{\beta}; \boldsymbol{\alpha}, \boldsymbol{\beta})|}{\binom{m-2l}{k-l}\binom{m}{2l}\binom{2l}{l}} \tag{2.3}$$

Proof. Let $\Sigma_{\mathbf{i}}$ denote the summation over all $\mathbf{i} = (i_1, i_2, \ldots, i_k)$ with $1 \leqslant i_1 < i_2 < \ldots < i_k \leqslant m$ and let $R_i = RB_0R'$, then

$$Ea_k(A, B) = Ea_k(D_a, B_0) = \Sigma_{\mathbf{i}} \left(\prod_{t=1}^{k} a_{i_t} \right) E|R_1(\mathbf{i}; \mathbf{i})| \tag{2.4}$$

Observe that

$$|R_1(\mathbf{i}; \mathbf{i})| = \Sigma_{\mathbf{j}} \Sigma_{\boldsymbol{\delta}} |B_0(\mathbf{j}; \boldsymbol{\delta})| \, |R(\mathbf{i}; \mathbf{j}| \, |R(\mathbf{i}; \boldsymbol{\delta})|$$

with $\mathbf{j} = (j_1, \ldots, j_k)$, $1 \leqslant j_1 < \ldots < j_k \leqslant m$; $\boldsymbol{\delta} = (\delta_1, \ldots, \delta_k)$ and $1 \leqslant \delta_1 < \delta_2 < \ldots < \delta_k \leqslant m$. Let

$$f(\mathbf{i}; \mathbf{j}; \boldsymbol{\delta}) = E\{|R(\mathbf{i}; \mathbf{j})| \, |R(\mathbf{i}; \boldsymbol{\delta})|\} \tag{2.5}$$

Then,

$$Ea_k(A, B) = \Sigma_{\mathbf{i}} \Sigma_{\mathbf{j}} \Sigma_{\boldsymbol{\delta}} \left(\prod_{t=1}^{k} a_{i_t} \right) |B_0(\mathbf{j}; \boldsymbol{\delta})| f(\mathbf{i}; \mathbf{j}; \boldsymbol{\delta}) \tag{2.6}$$

Given a particular $\mathbf{i}$, $\mathbf{j}$ and $\boldsymbol{\delta}$, assume that α_1, $\alpha_1 + \alpha_2$, $\alpha_1 + \alpha_3$ and $\alpha_1 + \alpha_4$ be respectively the common elements between (i) $\mathbf{i}, \mathbf{j}, \boldsymbol{\delta}$, (ii) $\mathbf{i}, \mathbf{j}$, (iii) $\mathbf{i}, \boldsymbol{\delta}$ and (iv) $\mathbf{j}, \boldsymbol{\delta}$. Let $k = (\alpha_1 + \alpha_2 + \alpha_3 + \alpha_5) = (\alpha_1 + \alpha_2 + \alpha_4 + \alpha_6) = (\alpha_1 + \alpha_3 + \alpha_4 + \alpha_7)$ and $\alpha_8 = m - \sum_{t=1}^{7} \alpha_t$. Since $ARA' \overset{d}{=} R$ for all permutation matrices, we shall write $U = (U_1, U_2, U_3, U_5, U_4, U_6, U_7, U_i)$ with $U_i \in \mathcal{R}^{p \times \alpha_i}$ for $i = 1, \ldots, 8$. If $U_i = U_i \Delta_i$ being an $\alpha_i \times \alpha_i$ orthogonal matrices for $i = 1, 2, \ldots, 8$, then

$$E|R(\mathbf{i}; \mathbf{j})| \, |R(\mathbf{i}; \boldsymbol{\delta})| = |\Delta_2| \, |\Delta_3| \, |\Delta_6| \, |\Delta_7| E|R(\mathbf{i}, \mathbf{j})| \, |R\{\mathbf{i}; \boldsymbol{\delta})|$$

and if $\alpha_2 > 0$ or $\alpha_3 > 0$ or $\alpha_6 > 0$ or $\alpha_7 > 0$, then choosing orthogonal matrices so that $|\Delta_2| \, |\Delta_3| \, |\Delta_6| \, |\Delta_7| = -1$ for all these cases. Then, for these situations, $f(\mathbf{i}; \mathbf{j}; \boldsymbol{\delta}) = 0$. Hence, we have to consider the situation : α_1 elements are common between $\mathbf{i}$, $\mathbf{j}$ and $\boldsymbol{\delta}$ while $\mathbf{j} = \boldsymbol{\delta}$. Hence, we can rewrite (2.6) as

$$Ea_k(A, B) = \sum_{l}^{*} \left(\prod_{t=1}^{k-l} a_{i_t} \right) \left(\prod_{t=1}^{l} a_{B_t} \right) |B_0(\mathbf{i}, \boldsymbol{\gamma}; \mathbf{i}, \boldsymbol{\gamma})| f_l \tag{2.7}$$

where,

$$f_l = E \begin{vmatrix} R_{11} & R_{12} \\ R_{31} & R_{32} \end{vmatrix}^2$$

R_{11}, R_{12}, R_{31} and R_{32} being respectively $(k-l) \times (k-l)$, $(k-l) \times l$, $l \times (k-l)$ and $l \times l$ matrices, $\mathbf{i} = (i_1, i_2, \ldots, i_{k-l})$, $i_1 < i_2 < \ldots < i_{k-l}$, $\boldsymbol{\gamma} = (\gamma_1, \ldots, \gamma_l)$, $\gamma_1 < \gamma_2 < \ldots < \gamma_l$ and $\beta_1 < \beta_2 < \ldots < \beta_l$.

Observe that

$$R_{11}, \; R_{22\cdot1} = R_{22} - R_{12}' R_{11}^{-1} R_{12}, \; R_{33\cdot12} = R_{33} - (R_{31}, R_{32}) \begin{pmatrix} R_{11} & R_{12} \\ R_{21} & R_{22} \end{pmatrix}^{-1} \begin{pmatrix} R_{13} \\ R_{23} \end{pmatrix}$$

and

$$U = (I - R_{33\cdot12})^{1/2} (R_{32} - R_{31} R_{11}^{-1} R_{12})(R_{22\cdot1} - R_{22\cdot1}^2)^{-1/2}$$

are independently distributed as

$$R_{11} \sim \beta_{k-l}\left(\tfrac{1}{2}p, \frac{m-p}{2}\right), \; R_{22.1} \sim \beta_l\left(\frac{p-k+1}{2}, \frac{m-p}{2}\right),$$

$$R_{33.12} \sim \beta_l(\tfrac{1}{2}(p-k), \tfrac{1}{2}(m-p))$$

and the density of U is constant

$$|I_1 - UU'|^{(1/2)(m-p-2l-1)} \text{ or } UU' \sim \beta_l(\tfrac{1}{2}l, \tfrac{1}{2}(m-p-l)).$$

Then

$$f_l = E|R_{11}|^2 E|R_{32-1}|^2$$
$$= E|R_{11}|^2 E|UU'|E|I - R_{33.12}|(E|R_{22.1}|\,|I - R_{22.1}|)$$

$$= \frac{\beta_{k-l}\left(\tfrac{1}{2}p+2, \frac{m-p}{2}\right)B_l\left(\tfrac{1}{2}l+1, \frac{m-p-l}{2}\right)B_l\left(\frac{p-k}{2}, \frac{m-p}{2}+1\right) \times B_l\left(\frac{p+l-k}{2}+1, \frac{m-p}{2}+1\right)}{B_{k-l}\left(\tfrac{1}{2}p, \frac{m-p}{2}\right)B_l\left(\tfrac{1}{2}l, \frac{m-p-l}{2}\right)B_l\left(\frac{p-k}{2}, \frac{m-p}{2}\right) \times B_l\left(\frac{p+l-k}{2}, \frac{m-p}{2}\right)}$$

$$= \binom{p}{k}\binom{p+2}{p-k+l+2}\binom{m-p}{l}\Big/\binom{k}{l}\binom{m-k}{l}\binom{m}{k}\binom{m+2}{k}$$

Now, using this in (2.7) and after some simplifications, the proof is complete.

Corollary 1. In Theorem 1, let B_0 be diagonal matrix

$$D_b = \text{diag}\,(b_1, b_2, \ldots, b_m) \text{ or } AB = BA.$$

Then

$$Ea_k(A, B) = \binom{p}{k} \sum_{l=1}^{k} W_l \bar{A}_l(A, B)$$

where $\bar{A}_l(A, B) = \Sigma_l^*\left(\prod_{t=1}^{k-l} a_{\alpha_t} b_{\alpha_t}\right)\left(\prod_{t=1}^{l} a_{\gamma t}\right)\left(\prod_{t=1}^{l} b_{\beta_t}\right)\Big/\binom{m-2l}{k-l}\binom{m}{2l}\binom{2l}{l}$ and W_l is defined in (2.2).

Corollary 2. In Theorem 1, if $A = \Sigma_1^{-1}$ and $B = \Sigma_1$ with eigen values of B as $\lambda_1 \geqslant \lambda_2 \geqslant \ldots \geqslant \lambda_m > 0$, then average efficiency is

$$Ee_k = Ea_k(\Sigma_1^{-1}, \Sigma_1)\Big/\binom{p}{k} = \sum_{l=1}^{k} W_l \bar{A}_l$$

where W_l is defined in (2.2) and

$$\bar{A}_l = \Sigma_l^{**}\{(\lambda_{\alpha_1}\lambda_{\alpha_2}\ldots\lambda_{\alpha_l})/(\lambda_{\beta_1}\lambda_{\beta_2}\ldots\lambda_{\beta_l})\}\Big/\binom{m}{2l}\binom{2l}{l}$$

with Σ_l^{**} being the summation over all possible combinations of $2l$ elements $\{\alpha_1, \ldots, \alpha_l; \beta_1, \ldots, \beta_l\}$ out of $1, 2, \ldots, m$ with $1 \leqslant \alpha_1 < \ldots < \alpha_l \leqslant m$ and $1 \leqslant \beta_1 < \beta_2 < \ldots < \beta_l \leqslant m$. When $X_1 = 0$ and $p = k$, we obtain the result of Iwasaki (1985).

3.3. TESTING OF LINEAR HYPOTHESIS

Let X be a set of $p \times n$ observations such that

$$X = \xi C' + \epsilon$$

where ϵ has a right spherical distribution, $\xi(\epsilon \mathcal{R}^{p\times c})$ is unknown and $C(\epsilon \mathcal{R}^{n\times c})$ is a known matrix of rank C. Suppose we want to test $H_0(\xi B' = 0)$ against $H(\xi B \neq 0)$ where $B(\epsilon R^{c\times b})$ is a known matrix of rank b. Let

$$C(C'C)^{-1}C' = L_1L_1' \text{ with } L_1'L_1 = I_c$$

where $C = L_1L_1'C$ and $L_1'C$ is a nonsingular matrix. Let

$$(C'L_1)^{-1}B'(B(C'C)^{-1}B')^{-1}B(L_1'C)^{-1} = B_1B_1' \text{ with } B_1'B_1 = I_b.$$

Further, let $B_0 = (B_1, B_2)$ and $L_0 = (L, L_1)$ be orthogonal matrices. Then, $T = L_0\begin{pmatrix} I_c & 0 \\ 0 & B_0 \end{pmatrix}$ is an orthogonal $n \times n$ matrix, and

$$\epsilon \overset{d}{=} \epsilon T = (X - \xi C')T = (X_1, -\xi C'L_1B_1 + X_2, -\xi C'L_1B_2 + X_3)$$

where $XL_1B_0 = (X_2, X_3)$, $XL = \epsilon L = X_1$ and under H_0,

$$\xi C'L_1BB_1' = \xi B'[B(C'C)^{-1}B']^{-1}B(L_1'C)^{-1} = 0 \text{ or } \xi C'L_1B_1 = 0.$$

Hence, under H_0,

$$(X_1, X_2) \text{ has a right spherical distribution} \tag{3.1}$$

and

$$(X_1, X_2) \overset{d}{=} S^{1/2}U = S^{1/2}(U_1, U_2) \tag{3.2}$$

where $S^{1/2}$ is a symmetric square root of $S = X_1X_1' + X_2X_2' = XAX' + XQX'$ with $A = I - C(C'C)^{-1}C'$ and $Q = C(C'C)^{-1}B'[B(C'C)^{-1}B']^{-1}B[C'C)^{-1}C']$, $S^{1/2}$ and U are independently distributed under H_0 and U is uniformly distributed over $O(p, n - c + b)$. Using Khatri's result (1970), the density of U_2 with $n \geqslant c + p$ is given by

$$\frac{\Gamma_p\left(\frac{n - c + b}{2}\right)}{\pi^{pb/2}\Gamma_p\left(\frac{n - c}{2}\right)} |I_p - U_2U_2'|^{(1/2)(n-c-p-1)} \tag{3.3}$$

for all $U_2 \in \mathcal{R}^{p\times b}$ such that $0 < U_2U_2' < I_p$. The result (3.3) is true under $H_0(\xi B' = 0)$. Hence, a function of the eigen values of U_2U_2' (or $U_2'U_2$ or $(XAX' + XOX')^{-1}XOX'$) can be used as a test statistic and under H_0, its distribution can be obtained from (3.3). Further, this test can be obtained for normal population and thus, it is null robust in the sense that it is applicable to a wider class of distributions. For some conditions on the density function of ϵ, it can be shown that the test is unbiased and for $p = 1$, it becomes uniformly most powerful invariant test [see, for example, Sinho and Drygas (1985) and Khatri (1984)].

3.4. EXACT MOMENTS OF $R = (R_{ij})$

For the linear model: $X = \xi C' + \epsilon$, it has been shown in section 3.3 that

$X_1 = XL$ has a right spherical distribution where $A = LL' = I - C(C'C)^{-1}C'$ and $L'L = I_{n-c}$. Then, the sample inner product between two observed points $\mathbf{x}_i$ and $\mathbf{x}_j$ from their estimate values is

$$R_{ij} = (\mathbf{x}_i - \hat{\mathbf{x}}_i)S^{-1}(\mathbf{x}_j - \hat{\mathbf{x}}_j) \text{ for } i, j = 1, 2, \ldots, n$$

where

$$X = (\mathbf{x}_1, \mathbf{x}_2, \ldots, \mathbf{x}_n),\ \hat{X} = XC(C'C)^{-1}C' = (\hat{\mathbf{x}}_1, \hat{\mathbf{x}}_2, \ldots, \hat{\mathbf{x}}_n),$$

and

$$S = \sum_{i=1}^{n} (\mathbf{x}_i - \hat{\mathbf{x}}_i)(\mathbf{x}_i - \hat{\mathbf{x}}_i)'/n.$$ Noting $X - \hat{X} = X_1L'$, we see that

$$R = (R_{ij}) = nLX_1'(X_1X_1')^{-1}X_1L' = nLR_0L' \tag{4.1}$$

where $R_0 = X_1'(X_1X_1')^{-1}X_1 = U'U \sim \beta_{n-c}[\frac{1}{2}p, \frac{1}{2}[n - c - p)]$ or U is uniformly distributed over $O(p, n - c)$. Then, using the results of James (1964) mentioned in (1.2), the moment generating function of $R[=(R_{ij})]$ is given by

$$M(\theta) = E \exp(tr\theta R) = E \exp(ntrL'\theta LU'U)$$

$$= \sum_{k=0}^{\infty} \Sigma_\kappa\, n^k(\tfrac{1}{2}p)_\kappa C_\kappa(\theta A)\Big/\left(\frac{n-c}{2}\right)_\kappa \tag{4.2}$$

where $(a)_\kappa = \prod_{j=1}^{p} \left(a - \frac{j-1}{2}\right)_{k_j}$, $(y)_\kappa = y(y+1)\ldots(y+k-1)$ for $k > 0$ and $(y)_0 = 1$, $\kappa = \{k_1, \ldots, k_p\}$ is a partition of κ into p parts and θ is any symmetric matrix. Let $A = LL' = I - C(C'C)^{-1}C' = (a_{ij})$. Equating the necessary coefficients of θ_{ij}'s from (4.2), we get the exact moments of R_{ij}'s. For the k-th order moments, we have

$$E(tr\theta R)^k = \Sigma_\kappa\, n^k(\tfrac{1}{2}p)_\kappa C_\kappa(\theta A)\Big/\left(\frac{n-c}{2}\right)_\kappa \tag{4.3}$$

and

$$ER_{ii}^k = n^k(\tfrac{1}{2}p)_k a_{ii}^k\Big/\left(\frac{n-c}{2}\right)_k \quad \text{for } k = 1, 2, \ldots \text{ and } i = 1, 2, \ldots, n \tag{4.4}$$

while, for $i \neq j$, $i, j = 1, 2, \ldots, n$.

$$ER_{ij}^k = \Sigma_\kappa\, n^k(\tfrac{1}{2}p)_\kappa C_\kappa(A_{ij})\Big/\left(\frac{n-c}{2}\right)_\kappa \tag{4.5}$$

where, $A_{ij} = \frac{1}{2}\begin{pmatrix} a_{ij} & a_{jj} \\ a_{ii} & a_{ij} \end{pmatrix}$, $k = \{k_1, k_2\}$, $k_1 \geqslant k_2 \geqslant 0$, $k = k_1 + k_2$, and using a result of James (1964),

$$C_\kappa(A_{ij}) = \frac{2^{2k_1}(k!)(k_1!)(2k_1 - 2k_2 + 1)}{(2k_1 + 1)!\, k_2!} \sum_{\nu_1 + 2\nu_2 = k_1 - k_2} \times \frac{(-1)^{\nu_2}(\frac{1}{2})_{\nu_1+\nu_2}}{\nu_1!\, \nu_2!} a_{ij}^{\nu_1}(|A_{ij}|)^{\nu_2+k_2} \tag{4.6}$$

Using (4.4) and (4.5), the first moments of

$$b_{1p} = \sum_{i,j=1}^{n} R_{ij}^3/n^2 \text{ and } b_{2p} = \sum_{i=1}^{n} R_{ii}^2/n^2 \tag{4.7}$$

are given by

$$Eb_{1p} = f_{11}(A)\,\frac{n^3p(p+2)(p+4)}{(n-c)(n-c+2)(n-c+4)}$$

$$+ \quad (A)\,\frac{3n^3p(p+2)(n-c-p)}{(n-c-1)(n-c)(n-c+2)(n-c+4)} \tag{4.8}$$

and

$$Eb_{2p} = f_{21}(A)n^2p(p+2)/(n-c)(n-c+2) \tag{4.9}$$

where,

$$f_{11}(A) = \left(\sum_{i,j=1}^{n} a_{ij}^3\right)\Big/ n^2,$$

$$f_{12}(A) = \sum_{i,j=1}^{n} a_{ij}(a_{ii}a_{jj} - a_{ij}^2)/n^2$$

and

$$f_{21}(A) = \left(\sum_{i=1}^{n} a_{ii}^2\right)\Big/ n.$$

For the second moment of b_{2p} we require ER_{ii}^4 and $ER_{ii}^2R_{jj}^2$ for $i \neq j$. ER_{ii}^4 is given by (4.4) and to obtain $E(R_{ii}^2R_{jj}^2)$, we use

$$E(\theta_{ii}R_{ii} + \theta_{jj}R_{jj})^k = \Sigma_\kappa\, n^k(\tfrac{1}{2}p)_\kappa C_\kappa((\theta A)_{ij})\Big/\left(\frac{n-c}{2}\right)_\kappa$$

with $\kappa = \{k_1, k_2\}$, $k_1 \geqslant k_2 \geqslant 0$ and $k_1 + k_2 = k$, where

$$(\theta A)_{ij} = \begin{pmatrix} \theta_{ii}a_{ii} & \theta_{ii}a_{ij} \\ \theta_{jj}a_{ij} & \theta_{jj}a_{jj} \end{pmatrix},$$

$$C_\kappa(\theta A)_{ij} = \frac{2^{2k_1}(k!)(k_1!)(2k_1 - 2k_2 + 1)}{(2k_1+1)!\,k_2!} \sum_{\nu_1+2\nu_2=k_1-k_2} \frac{(-1)^{\nu_2}(\frac{1}{2})\nu_1 + \nu_2}{\nu_1!\,\nu_2!}$$

$$\times \frac{(-1)^{\nu}(\frac{1}{2})_{\nu_1+\nu_2}}{\nu_1!\,\nu_2!}(\theta_{ii}a_{ii} + \theta_{jj}a_{jj})^{\nu_1}|(\theta A)_{ij}|^{\nu_2+k_2} \tag{4.10}$$

Equating the coefficients of $\theta_{ii}^2\theta_{jj}^2$ with $k = 4$, we get

$$E(R_{ii}^2R_{jj}^2) = \frac{n^4p(p+2)(p+4)(p+6)}{(n-c)(n-c+2)(n-c+4)(n-c+6)}\,[a_{ii}^2a_{jj}^2 - \tfrac{8}{7}a_{ii}a_{jj}(a_{ii}a_{jj} - a_{ij})$$

$$+ \tfrac{8}{35}(a_{ii}a_{jj} - a_{ij}^2)^2] + \frac{n^4(p-1)p(p+2)(p+4)}{(n-c-1)(n-c)(n-c+2)(n-c+4)}$$

$$\times [\tfrac{8}{7}a_{ii}a_{jj}(a_{ii}a_{jj} - a_{ij}^2) - \tfrac{16}{21}(a_{ii}a_{jj} - a_{ij}^2)^2]$$

$$+ \frac{n^4(p-3)(p-1)p(p+2)}{(n-c-3)(n-c-1)(n-c)(n-c+2)}\left(\frac{8}{15}\right)(a_{ii}a_{jj} - a_{ij}^2)^2$$

and hence,

$$E(R_{ii}^2R_{ij}^2) = g_1(p,n)a_{ii}^2a_{jj}^2 + g_2(p,n)a_{ii}a_{jj}(a_{ii}a_{jj} - a_{ij}^2) + g_3(p,n)(a_{ii}a_{jj} - a_{ij}^2)^2 \tag{4.11}$$

where,

$$g_1(p,n) = n^4p(p+2)(p+4)(p+6)/(n-c)(n-c+2)(n-c+4)(n-c+6) \tag{4.12}$$

$$g_2(p,n) = 8n^4p(p+2)(p+4)(n-c-p)/(n-c-1)(n-c)$$
$$\times (n-c+2)(n-c+4)(n-c+6) \tag{4.13}$$

and

$$g_3(p, n) = \frac{8n^4p(p+2)(n-c-p)(-4(n-c)p+19(n-c)-51p)}{(n-c-3)(n-c-1)(n-c)(n-c+2)(n-c+4)(n-c+6)} \quad (4.14)$$

Thus, taking

$$f_{22}(A) = [f_{21}(A)]^2 = \left(\sum_{i=1}^{n} a_{ii}^2\right)^2 \Big/ n^2,$$

$$f_{23}(A) = \sum_{i,j} a_{ii}a_{jj}(a_{ii}a_{jj} - a_{ij}^2)/n^2 = (f_{21}(A))^2 - \sum_{i,j} a_{ii}a_{jj}a_{ij}^2/n^2$$

and

$$f_{24}(A) = \sum_{i,j} (a_{ii}a_{jj} - a_{ij}^2)^2/n^2 = [f_{21}(A)]^2 - 2(\sum_{i,j} a_{ii}a_{jj}a_{ij}^2/n^2) + \sum_{i,j} a_{ij}^4/n^2,$$

we get

$$Eb_{2p}^2 = g_1(p, n)f_{22}(A) + g_2(p, n)f_{23}(A) + g_3(p, n)f_{24}(A) \quad (4.15)$$

The exact second moment of b_{1p} is very complicated and the details to establish it explicitly using the above type of arguments are omitted. The asymptotic distribution of b_{1p} and b_{2p} will be similar to those of Mardia (1970, 1974), but this is not mentioned explicitly.

REFERENCES

1. Dawid, A. P. (1977): Spherical matrix distributions and a multivariate model, *J. Royal Statist. Soc.* Series **B**, **39**, 254–261.
2. Iwasaki, Manabu (1985): Mean efficiency of least squares estimator of regression coefficients, to appear in *Journal of the Japan Statistical Society.*
3. James, A. T. (1964): Distribution of matrix variates and latent roots derived from normal samples, *Ann. Math. Statist.*, **35**, 475–501.
4. Khatri, C. G. (1970): A note on Mitra's paper, "A density free approach to the matrix variate beta distribution", *Sankhya*, Series **A**, **32**, 311–318.
5. ——— (1984a): Some remarks on the spherical distributions and linear models; *Proceedings of the International Conference on "Linear Statistical Inference"* at Poznan (Poland) (edited by T. Calinski and W. Klonecki).
6. ——— (1984b). Robustness study for a linear growth curve (sent for publication to *J. Multivariate Analysis*).
7. Mardia, K. V. (1970): Measures of Multivariate skewness and kurtosis with applications, *Biometrika*, **47**, 519–530.
8. ——— (1974): Applications of some measures of multivariate skewness and kurtosis for testing normality and robustness study, *Sankhya* series **B**, **36**, 115–128.
9. Mitra, S. K. (1970): A density free approach to matrix variate beta distribution, *Sankhya* series **A**, **32**, 81–88.
10. Sinha, B. K. and Drygas, H. (1985): Robustness in linear models, Proceedings of the first International Tampere Seminar on linear statistical models and their applications (edited by Tarmo Pukkila and Simo Puntanen), pp. 143–160.

4

Strong Consistency of Least Squares Estimates in Time Series

B. R. Bhat and K. Suresh Chandra
Karnatak University and Madras University

4.1 INTRODUTION

Let $\{x(t),\ t = 1, 2, \ldots\}$ be a time series generated by the model

$$x(t) = \theta_1 a_1(t) + \theta_2 a_2(t) + \ldots + \theta_k a_k(t) + \epsilon(t),\ (t = 1, 2, \ldots) \quad (1.1)$$

where $\theta = (\theta_1, \theta_2, \ldots, \theta_k)'$ is the vector of unknown parameters, $\{a_i(t)\}$ are sequences whose values are known at time t, some of which may be lagged variables of $x(t)$, some may be polynomials in t and some may be exogenous variables $(i = 1, 2, \ldots, k)$. $\{\epsilon(t), \mathcal{F}_t\}$ is assumed to be a martingale difference sequence such that

(i) $E(\epsilon(t) \mid \mathcal{F}_{t-1}) = 0\ (t \geqslant 1)$,

(ii) $\text{Var}\ (\epsilon(t) \mid \mathcal{F}_{t-1}) = \sigma^2$ (1.2)

where $\mathcal{F}_t$ is the σ-field generated by $\{x(1), \ldots, x(t)\}$. Vectorially (1.1) may be written as

$$x(t) = A'(t)\theta + \epsilon(t),$$

where $A'(t) = (a_1(t), a_2(t), \ldots, a_k(t))$ and $\theta = (\theta_1, \ldots, \theta_k)'$. If $X_n = (x_{(1)}, \ldots, x_{(n)})'$, $\epsilon_n = (\epsilon_{(1)}, \epsilon_{(2)}, \ldots, \epsilon_{(n)})'$ and A_n is the $n \times k$ matrix whose t-th row is $A'(t)$, (1.1) can be expressed as

$$X_n = A_n\theta + \epsilon_n. \quad (1.3)$$

We assume without loss of generality that $x(t) = 0$ for $t \leqslant 0$ which implies that $\epsilon(t) = 0$ for $t \leqslant 0$. Even when some of the regressors are lagged variables, $x(t)$ will be a function of $\epsilon_{(1)}, \epsilon_{(2)}, \ldots, \epsilon_{(t)}$ and consequently $\mathcal{F}_t$ may be taken to be the σ-field generated by $\epsilon_{(1)}, \ldots, \epsilon_{(t)}$. We also assume that exogenous variables are not present in the model.

The least squares (l.s.) estimate $\widehat{\theta}_n$ of θ after observing $x_{(1)}, \ldots, x_{(n)}$ is the solution of the normal equations

$$A_n' A_n \widehat{\theta}_n = A_n' X_n. \quad (1.4)$$

Hence forward we shall omit the suffix n when there is no ambiguity.

Using (1.3), (1.4) can also be written as

$$A'A(\widehat{\theta} - \theta) = A'\epsilon. \tag{1.5}$$

When $\{\epsilon(t), \mathcal{F}_t\}$ is a martingale difference sequence, $\sum_{t=1}^{n} a_i(t)\epsilon(t)$ is a martingale. Since $a_i(t)$ is $\mathcal{F}_{t-1}$ measurable, the sum of conditional variances of this martingale is given by $\sigma^2 a_{ii}(n)$, where

$$a_{ii}(n) = \sum_{i}^{n} a_i^2(t), \ (i = 1, 2, \ldots, k) \tag{1.6}$$

is the i-th diagonal element of $A'A$. Thus, $A'\epsilon$ is a vector of martingales with the sum of conditional covariances matrix as $\sigma^2 A'A$ [see Lai and Wei, (1982)]. Asymptotic behaviour of $(\widehat{\theta} - \theta)$ is related to that of $A'\epsilon$. This fact will be exploited in the coming sections for establishing strong consistency of $\widehat{\theta}$.

Strong consistency of $\widehat{\theta}$ is of interest to engineers and social scientists, because they may have only a single sequence of observations from which they can draw inferences on θ [see Lai and Wei (1983)]. In the case of multiple regression on exogeneous variables, strong consistency of l.s. estimates of regression coefficients was discussed by Bhat (1982). The same was established by Lai and Wei (1982) when some of the $a_i(t)$'s are non-random and some others are random, on assuming condition (1.2)–(i) but not (1.2)–(ii). They have also stated that in the case where $a_i(t)$'s are non-random, $a_{ii}(n) \to \infty$ for all i, is necessary and sufficient for strong consistency, if (1.2) is satisfied. In this paper, under a few additional assumptions, we establish strong consistency even when some of the $a_i(t)$'s are $\mathcal{F}_{t-1}$ measurable random variables.

4.2. THE MAIN THEOREM

Before proceeding to establish our main result, we state the following lemma which is usually called the law of large numbers for martingales [see Hall and Heyde, pages 36, 158, (1980)].

Lemma 1. Let $\{Y_i, \mathcal{F}_i\}$ be a sequence of martingale differences with $E(Y_i \mid \mathcal{F}_{i-1}) = 0$, Var $(Y_i \mid \mathcal{F}_{i-1}) = \sigma_i^2$, $(i = 1, 2, \ldots)$. Let $s_n^2 = \sum_{1}^{n} \sigma_i^2$, then on the set $[s_n^2 \uparrow \infty]$, $\sum_{1}^{n} Y_i/s_n^2$ converges to zero a.s. (*)

Since the i-th element of the right side of (1.5) is a martingale, with $a_{ii}(n)$ defined by (1.6) as its conditional variance, by Lemma 1, on $[a_{ii}(n) \uparrow \infty]$,

$$\sum_{1}^{n} a_i(t) \, \epsilon \, (t)/a_{ii}(n) \to 0 \quad \text{a.s.} \tag{2.1}$$

Dividing the i-th element of (1.5) by $a_{ii}(n)$, the left side of (1.5) will be $\Delta_n^{-1}A'A(\widehat{\theta} - \theta)$, where Δ_n is the diagonal matrix with its i-th element

$a_{ii}(n)$ $(i = 1, 2, \ldots, k)$, and allowing $n \to \infty$, the right side tends to zero on the set $[a_{ii}(n) \uparrow \infty, i = 1, 2, \ldots, k]$, i.e., on the set $[(A'A)^{-1} \to 0]$. Then (2.1) is equivalent to

$$(\widehat{\theta}_{in} - \theta_i) + \sum_{j \neq i} (a_{ij}/a_{ii})(\widehat{\theta}_{jn} - \theta_j) \to 0 \quad \text{a.s.} \tag{2.2}$$

This would ensure that $(\widehat{\theta}_n - \theta)$ would tend to zero a.s., if $\lim \Delta_n^{-1}A'A$ is nonsingular. Hence we have the following theorem:

Theorem 2.1. For the model (1.3), the least squares estimator $\widehat{\theta}_n$ of θ given by (1.4) will be strongly consistent if $\Delta_n \to \infty$ and $\Delta_n^{-1}A'A$ converges to a non-singular matrix, where Δ_n is the diagonal matrix with its i-th diagonal element as $a_{ii}(n)$ given in (1.6). (*)

Note that $\Delta_n^{-1}A'A$ is a matrix with diagonal element unity and (i, j)-th element given by $a_{ij}(n)/a_{ii}(n)$, where

$$a_{ij}(n) = \sum_{t=1}^{n} a_i(t)a_j(t).$$

If $\lim [a_{ij}(n)/a_{ii}(n)]$ exists, it may be 0, $\pm\infty$ or finite. If it is zero for all $j \neq i$, then from (2.2) we have consistency of $\widehat{\theta}_{in}$.

If $|a_{ij}(n)/a_{ii}(n)| \to \infty$ for a certain j and i, $|\widehat{\theta}_{jn} - \theta_j|$ must tend to zero in order to satisfy (2.2). This implies consistency of $\widehat{\theta}_{jn}$. These results are summarised in the following corollary:

Corollary 2.1. If the (i, j)-th element of $\Delta_n^{-1}A'A$ (viz., a_{ij}/a_{ii}) tends to infinity, then $\widehat{\theta}_{jn}$ is a consistent estimator of θ_j. If it tends to zero for all $j \neq i$, then $\widehat{\theta}_{in}$ is a consistent estimator of θ_i. (*)

Lai and Wei (1982) have quoted an example of a least squares estimator involving two parameters which is not consistent. There we see that $\lim \Delta_n^{-1}A'A$ is singular, one of the estimators is inconsistent and hence the other one is also inconsistent. In the remaining sections, we shall discuss some examples from the time series models.

4.3 CONSISTENCY IN *AR*(*K*) MODELS

In (1.1) suppose $a_i(t) = x(t - i)(i = 1, 2, \ldots, k)$, then we have an autoregressive model of order k without trend or constant term, and with a martingale difference error term, called $AR(k)$ model. Then (i, j)-th element of $A'A$ becomes

$$a_{ij}(n) = \sum_{t=1}^{n} x(t - i)x(t - j), \tag{3.1}$$

where we take $x(t - i) = 0$ for $i \geqslant t$. Divergence of $a_{ii}(n) = \sum_{1}^{n} x^2(t - i)$ to infinity for $i = 1, 2, \ldots, k$, is a necessary condition for consistency of $\widehat{\theta}$. We shall verify this and other conditions of the Theorem 2.1, and/or of the Corollary 2.1, under different cases.

Case 1: (Non-explosive case). Suppose, a priori, we know the θ_i's are such that the polynomial

$$P(z) = z^k - \theta_i z^{k-1} - \ldots - \theta_k, \tag{3.2}$$

has all its roots lying within the unit circle. Then $x(t)$ has the representation of the form

$$x(t) = \sum_{r=0}^{t-1} \lambda(r)\epsilon(t-r), (t \geqslant 1), \tag{3.3}$$

where $\lambda(r)$'s satisfy the recurrence relation

$$\lambda(r) - \theta_1\lambda(r-1) - \ldots - \theta_k \lambda(r-k) = 0 \ (r = 1, 2, \ldots), \tag{3.4}$$

[see Anderson, (1971)].

Define the covariance function

$$Q(s) = Q(-s) = \sigma^2 \sum_{r=0}^{\infty} \lambda(r)\lambda(r+|s|), \tag{3.5}$$

of the $x(t)$ process which is asymptotically stationary.

Therefore, by ergodic theorem

$$\frac{1}{n} \sum_t x(t)x(t+s) \to Q(s), \text{ a.s. } (s \geqslant 0). \tag{3.6}$$

Since $Q(s)$ is finite and nonzero, $a_{ij}(n) \uparrow \infty$ a.s: for all i.

Moreover, since $a_{ii}(n)$ and $a_{jj}(n)$ differ only by a finite number of terms,

$$\lim_n \frac{a_{ij}(n)}{a_{ii}(n)} = \lim_n \frac{a_{ij}(n)}{\sqrt{(a_{ii}(n)a_{jj}(n))}} = \frac{Q(|i-j|)}{Q(0)} = \rho_{ij}, \tag{3.7}$$

where (ρ_{ij}) is the matrix of order k of the serial correlations of the $x(t)$ process. Since model (1.1) implies that there exists no exact linear relationship between k consecutive values of $x(t)$, (ρ_{ij}) is non-sigular. Therefore, by Theorem 2.1, least squares estimators are strongly consistent.

Case 2: (Explosive Case). Suppose the polynomial $P(z)$ defined by (3.2) has all its roots lying outside the unit circle.

Without loss of generality let us take them to be simple. Let z_1 be the root of $P(z)$ with largest modulus. As in Lai and Woi (1983), denoting

$$x_{n-1} = \begin{bmatrix} x(n-1) \\ x(n-2) \\ \vdots \\ \vdots \\ x(n-k) \end{bmatrix}, e_n = \begin{bmatrix} \epsilon(n) \\ 0 \\ \vdots \\ \vdots \\ 0 \end{bmatrix}, B = \left[\begin{array}{c|c} \theta_1\theta_2\theta_{k-1} & \theta_k \\ \hline I_{k-1} & 0 \end{array}\right], \tag{3.8}$$

we have

$$\begin{aligned} x_n &= Bx_{n-1} + e_n, \\ &= B^{n-k}x_k + B^{n-k-1}e_{k+1} + \ldots + Be_{n-1} + e_n. \end{aligned} \tag{3.9}$$

Therefore,

$$B^{-n+k}x_n = x_k + B^{-1}e_{k+1} + \ldots + B^{-n+k+1}e_{n-1} + B^{-n+k}e_n, \tag{3.10}$$

where $x_k = 0$. Since the right side of (3.10) is a martingale whose variance is bounded and non-zero, it converges a.s. to a random vector $\mathbf{Z} \neq 0$.

Since the roots of $P(z)$ are the latent roots of B, using spectral decomposition of B, or otherwise, we see that B^n increases as fast as Z_1^n. If W is the latent matrix associated with z_1

$$z_1^{-n} B^n \rightarrow W \tag{3.11}$$

and hence

$$z_1^{-n+k} x_n \rightarrow W \cdot \mathbf{Z} = U \text{ (say)} \neq 0, \text{ a.s.} \tag{3.12}$$

This implies that x_n increases exponentially fast and that, as $n \rightarrow \infty$,

$$z_1^{-2n} x_n x_n' \rightarrow z_1^{-2k} UU', \text{ a.s.} \tag{3.13}$$

By Toeplitz lemma (3.13) implies that

$$\sum_1^n x_j x_j' \Big/ \left(\sum_1^n z_1^{2j} \right) \rightarrow z_1^{-2k} UU' \neq 0, \text{ a.s.} \tag{3.14}$$

Hence $\sum_1^n x_j x_j' = A'A$ increases exponentially fast. In particular, all the diagonal elements of $A'A$ increase to infinity as fast as z_1^{2n}. From (3.12) we see that

$$a_{ij}(n)/a_{ii}(n) \rightarrow u_j/u_i \ (j \neq i), \text{ a.s.} \tag{3.15}$$

where $U' = (u_1, \ldots, u_k)$. Hence $\lim \Delta_n^{-1} A'A$ defined in section 4.2 is a singular matrix. Therefore we cannot claim consistency of each component of $\hat{\theta}_n$. We can only assert that there exists a linear function of $\hat{\theta}_{jn}$'s which consistently estimates the corresponding linear function of θ_j's. We also note that if there is only one parameter θ, it can be consistently estimated.

4.4. CONSISTENCY OF LEAST SQUARES ESTIMATOR USING AUXILIARY PROCESSES

A slight modification in the arguments given in Section 4.3 will provide consistent estimator for each one of the parameters, for the purely explosive case. Let $z_1, z_2, \ldots, z_k$, the roots of $p(z)$ defined in (3.2), have the placement

$$|z_1| > |z_2| > \ldots > |z_k| > 1 \tag{4.1}$$

We define the auxiliary processes $\{\{x_i(t)\}, i = 1, 2, \ldots, k\}$ using (4.1) given by

$$x_1(t) = x(t)$$

$$x_i(t) = x_{i-1}(t) - z_{i-1} x_{i-1}(t-1); \ i = 2, \ldots, k. \tag{4.2}$$

It can be verified that $[x_i(t), t \geqslant k + 2 - i]$ satisfy the stochastic difference equation

$$x_i(t) - \theta_{i,1} x_i(t-1) - \ldots - \theta_{i,k+1-i} x_i(t-k-1+i) = \epsilon(t) \tag{4.3}$$

where $\theta_{i,r}$ are functions of θ_i; $i = 1, \ldots, k$ and the root with largest modulus of the characteristic polynomial associated with (4.3), namely,

$$P_i(z) = z^{k+1-i} - \theta_{i,1} z^{k+1-i-1} - \ldots - \theta_{i,k+1-i} \tag{4.4}$$

is precisely z_i. As in (3.9) and (3.11), let us denote

$$x_{i,n-1} = \begin{bmatrix} x_i(n-1) \\ x_i(n-2) \\ \vdots \\ x_i(n-k+1-i) \end{bmatrix}; \; e_{i,n} = \begin{bmatrix} \epsilon(n) \\ 0 \\ \vdots \\ 0 \end{bmatrix}$$

$$B_i = \left[\begin{array}{c|c} \theta_{i1}, \ldots, \theta_{i,k+1-i-1} & \theta_{i,k+1-i} \\ \hline I_{k+1-i-1} & 0 \end{array}\right] \tag{4.5}$$

we have for $i = 1, \ldots, k$,

$$\begin{aligned} x_{i,n} &= B_i x_{i,n-1} + e_{i,n} \\ &= B_i^{n-(k+1-i)} x_{i,k+1-i} + B_i^{n-(k+1-i)-1} e_{i,k+1-i-1} + \ldots + B_i e_{i,n-1} + e_{i,n} \end{aligned} \tag{4.6}$$

Therefore, as in Section 4.3,

$$B_i^{-n+(k+1-i)} x_{i,n} = x_{i,k+1-i} + B_i^{-1} e_{i,k+1-i+1} + \ldots + B_i^{-n+k+2-i} e_{i,n-1} + B_i^{-n+(k+1-i)} e_{i,n} \tag{4.7}$$

where $x_{i,k+1-i} = 0$. Since the right side of (4.7) is a martingale whose variance is bounded and non-zero, it converges almost surely to a random vector ξ_i, (say) ($\neq 0$).

Since the roots of $P_i(z)$ are the latent roots of B_i, we see, as before, that B_i^n increases as fast as z_i^n.

If W_i is the latent matrix associated with z_i

$$z_i^{-n} B_i^n \to W_i, \tag{4.8}$$

and hence for $i = 1, 2, \ldots, k$,

$$z_i^{-n+(k+1-i)} x_{i,n} \xrightarrow{\text{a.s.}} W_i \xi_i = \mathbf{U}_i \quad \text{(say)} \neq 0 \tag{4.9}$$

This implies that $x_{i,n}$ increases as fast as z_i^n as $n \to \infty$, and consequently, for $i, j = 1, 2, \ldots, k$, we have

$$\begin{aligned} z_i^{-2n} x_{i,n} x'_{,n} &\to z_i^{-2(k+1-i)} \mathbf{U}_i \mathbf{U}'_i, \quad \text{a.s.} \\ z_i^{-n} z_j^{-n} x_{i,n} x'_{j,n} &\to z_i^{-(k+1-i)} z_j^{-(k+1-j)} \mathbf{U}_i \mathbf{U}'_j \quad \text{a.s.} \end{aligned} \tag{4.10}$$

By Toeplitz lemma it implies from (4.10) that for $i, j = 1, \ldots, k$,

$$\begin{aligned} \sum_{r=1}^{n} x_{i,r} x'_{,r} \Big/ \sum_{r=1}^{n} z_i^{2r} &\to z_i^{-2(k+1-i)} \mathbf{U}_i \mathbf{U}'_i \quad \text{a.s.} \\ \sum_{r=1}^{n} x_{i,r} x'_{j,r} \Big/ \sum_{r=1}^{n} z_i^r z_j^r &\to z_i^{-(k+1-i)} z_j^{-(k+1-j)} \mathbf{U}_i \mathbf{U}' \quad \text{a.s.} \end{aligned} \tag{4.11}$$

Let $R_{ij}(k)$ denote the elementary row operation of adding k-times j-th row of a matrix to its i-th row, and similarly let $C_{ij}(k)$ denote the corresponding operation relating to the columns of a matrix. The successive operations are effected in left to right order. Let us use the operator, [as done by Venkataraman (1968)]

$$\prod_{i=1}^{k-1} \prod_{j=1}^{i} R_{i-j+1,\, i-j+2}(-z_{k-i}) \prod_{i=1}^{k-1} \prod_{j=1}^{i} C_{i-j+1,\, i-j+2}(-z_{k-i}) \tag{4.12}$$

on the matrix $A'A$ to obtain the matrix $D = (d_{ij})$ where

$$d_{ij} = \sum_{t=1}^{n} x_{k-i+1}(t-i)x_{k-j+1}(t-j);\ i, j = 1, \ldots, k. \tag{4.13}$$

Operating only the row transformations defined in (4.12) on the vector $A'\epsilon$, we get the vector $d = (d_1, \ldots, d_k)'$ where

$$d_i = \sum_{t=1}^{n} x_{k-i+1}(t-i)\epsilon(t);\ i = 1, \ldots, k. \tag{4.14}$$

These modifications will transform the system of normal equations in (1.5), to

$$D(\hat{\eta} - \eta) = d$$

where the elements of the vector $(\hat{\eta} - \eta)$ are recursive linear combinations of the elements of $(\hat{\theta} - \theta)$. To be specific, when $k = 2$,

$$(\hat{\eta} - \eta) = \begin{bmatrix} (\hat{\theta}_1 - \theta_1) \\ z_1(\hat{\theta}_1 - \theta_1) + (\hat{\theta}_2 - \theta_2) \end{bmatrix} \tag{4.16}$$

and when $k = 3$

$$(\hat{\eta} - \eta) = \begin{bmatrix} (\hat{\theta}_1 - \theta_1) \\ -(z_1 + z_2)(\hat{\theta}_1 - \theta_1) + (\hat{\theta}_2 - \theta_2) \\ z_1^2(\hat{\theta}_1 - \theta_1) + z_1(\hat{\theta}_2 - \theta_2) + (\hat{\theta}_3 - \theta_3) \end{bmatrix} \tag{4.17}$$

We now note that $\{d_i\}$ is a martingale and d_{ii} is proportional to its conditional variance. From (4.10), it follows that $d_{ii} \uparrow \infty$, as $n \to \infty$, as fast as z_i^{2n}, so that an appeal to Lemma 1 will yield that

$$(d_{ii})^{-1} d_i \to 0 \quad \text{(a.s.)},\ i = 1, \ldots, k \tag{4.18}$$

Let $V=(v_1, \ldots, v_k)'$ where v_i is the $(k-i+1)$-st element of U_i; $i=1, \ldots, k$, defined in (4.9). Using V and denoting $\Delta_n(D) = \text{diag}\ (d_{ii};\ i = 1, \ldots, k)$ it can be easily verified from (4.10) and (4.11) that

$$\det\, [\Delta_n(D)]^{-1}D \to c \det \Delta \text{ (say)} \quad \text{a.s.} \tag{4.19}$$

where $c \neq 0$, and Δ has $(z_iz_j - 1)^{-1}/(z_i^2 - 1)^{-1/2}(z_j^2 - 1)^{-1/2}$ as its $(k - i + 1, k - j + 1)$-st element for $i, j = 1, \ldots, k$. Consequently by virtue of (4.1), Δ is non-singular.

This observation together with an appeal to Theorem 2.1 will enable us to conclude that $\hat{\eta}$, and hence $\hat{\theta}$ are strongly consistant for the respective parameter vectors. A similar result can also be obtained even when the roots of $P(z)$ are not distinct, following the arguments of Venkataraman (1973).

4.5 PARTIALLY EXPLOSIVE CASE

Suppose θ_i's are such that the polynomial $P(z)$ defined in (3.2) is of the form $P_1(z) \cdot P_2(z)$, where

$$P_1(z) = z^m - \beta_1 z^{m-1} - \ldots - \beta_m$$

$$P_2(z) = z^l - \gamma_1 z^{l-1} - \ldots - \gamma_l \tag{5.1}$$

We assume that the roots $\rho_1, \rho_2, \ldots, \rho_m$ of $P_1(z)$ are such that $|\rho_i| < 1$ and the roots $\phi_1, \phi_2, \ldots, \phi_l$ of $P_2(z)$ are such that $|\phi_j| > 1$. Then defining

$$y(t) = x(t) - \gamma_1 x(t-1) - \ldots - \gamma_l x(t-l), \tag{5.2}$$

we have

$$y(t) - \beta_1 \cdot y(t-1) - \ldots - \beta_m y(t-m) = \epsilon(t). \tag{5.3}$$

Since $y(t)$ process behaves like the $x(t)$ process of Case 1 of section 4.3 discussed earlier

$$y(t) = \sum_{r=0}^{t-1} h(r)\epsilon(t-r), \tag{5.4}$$

where $h(r)$'s satisfy the recurrence relation

$$h(r) - \beta_1 h(r-1) - \ldots - \beta_m h(r-m) = 0. \tag{5.5}$$

On closely following the arguments of Cases 1 and 2 of section 4.3, it can be verified that $\Delta_n^{-1} A'A$ defined in Theorem 2.1 will converge to a singular matrix under the present assumptions on the roots of $P(z)$. Therefore one cannot claim the consistency of each component of the least squares estimator $\hat{\theta}$. We can only assert that a linear combination of θ_i's can be consistently estimated.

However as in section 4.4, one can establish the consistency of $\hat{\theta}$ using auxiliary processes.

Let us assume that the roots $\phi_1, \phi_2, \ldots, \phi_l$ of $P_2(z)$ have the placement

$$|\phi_1| > |\phi_2| > \ldots > |\phi_l| > 1, \tag{5.6}$$

Defining, as before, the auxiliary processes

$$x_1(t) = x(t)$$

$$x_i(i) = x_{i-1}(t) - \phi_{i-1} x_{i-1}(t-1); \; i = 2, \ldots, l+1, \tag{5.7}$$

it can be verified that, for $i = 1, 2, \ldots, l$, $x_i(t)$ satisfy equations of the type (4.3) and $x_{l+1}(t)$ satisfies the stochastic difference equation

$$x_{l+1}(t) - \beta_1 x_{l+1}(t-1) - \ldots - \beta_m x_{l+1}(t-m) = \epsilon(t). \tag{5.8}$$

We note that $x_{l+1}(t)$ process behaves like $x(t)$ process discussed under

Case 1 of section 4.3; and consequently the results similar to those in (3.3), (3.6) and (3.7) hold for the $x_{l+1}(t)$ process.

As done in section 4.4, let us use the transformation

$$RC = \prod_{i=1}^{l} \prod_{j=1}^{i} R_{i-j+1,\, i-j+2}(-\phi_{l-1}) \prod_{i=1}^{l} \prod_{j=1}^{i} C_{i-j+1,\, i-j+2}(-\phi_{l-i}) \quad (5.9)$$

on the matrix $A'A$ in (1.5), and only the row transformations in (5.9) on the vector $A'\epsilon$ in (1.5) to obtain the system of equivalent equations

$$G(\hat{\alpha} - \alpha) = g. \quad (5.10)$$

It can be verified that the matrix G involves sums of squares and products of the auxiliary processes defined in (5.7), the vector g the sums of products of these processes and $\epsilon(t)$, and the vector $(\hat{\alpha} - \alpha)$ is a specific recursive linear combinations of $(\hat{\theta}_i - \theta_i)$, $i = 1, \ldots, k$. Specifically when $k = 2$, $l = 1$, we have

$$G = \begin{bmatrix} \Sigma x_2^2(t-1) & \Sigma x_2(t-1)x_1(t-2) \\ \Sigma x_1(t-1)x_2(t-2) & \Sigma x_1^2(t-2) \end{bmatrix};\; g = \begin{bmatrix} \Sigma x_2(t-1)\epsilon(t) \\ \Sigma x_1(t-2)\epsilon(t) \end{bmatrix}$$

$$(\hat{\alpha} - \alpha)' = [(\hat{\theta}_1 - \theta_1),\, \phi_1(\hat{\theta}_1 - \theta_1) + (\hat{\theta}_2 - \theta_2)]' \quad (5.11)$$

and for the case $k = 4$, $l = 2$

$$G = \begin{bmatrix} \Sigma x_3^2(t-1) & \Sigma x_3(t-1)x_3(t-2) & \Sigma x_3(t-1)x_2(t-3) & \Sigma x_3(t-1)x_1(t-4) \\ \Sigma x_3(t-2)x_3(t-1) & \Sigma x_3^2(t-2) & \Sigma x_3(t-2)x_2(t-3) & \Sigma x_3(t-2)x_1(t-4) \\ \Sigma x_2(t-3)x_3(t-1) & \Sigma x_2(t-3)x_3(t-2) & \Sigma x_2^2(t-3) & \Sigma x_2(t-3)x_1(t-4) \\ \Sigma x_1(t-4)x_3(t-1) & \Sigma x_1(t-4)x_3(t-2) & \Sigma x_1(t-4)x_2(t-3) & \Sigma x_1^2(t-4) \end{bmatrix},$$

$$g = \begin{bmatrix} \Sigma x_3(t-1)\;\; \epsilon(t) \\ \Sigma x_3(t-2)\;\; \epsilon(t) \\ \Sigma x_2(t-3)\;\; \epsilon(t) \\ \Sigma x_1(t-4)\;\; \epsilon(t) \end{bmatrix}, \text{ and}$$

$$(\hat{\alpha} - \alpha) = \begin{bmatrix} (\hat{\theta} - \theta_1) \\ (\phi_1 + \phi_2)(\hat{\theta}_1 - \theta_1) + (\hat{\theta}_2 - \theta_2) \\ ((\phi_1 + \phi_2)\phi_2 + \phi_1^2)(\hat{\theta}_1 - \theta_1) + (\phi_1 + \phi_2)(\hat{\theta}_2 - \theta_2) + (\hat{\theta}_3 - \theta_3) \\ \phi_1^3(\hat{\theta}_1 - \theta_1) + \phi_1^2(\hat{\theta}_2 - \theta_2) + \phi_1(\hat{\theta}_3 - \theta_3) + (\hat{\theta}_4 - \theta_4) \end{bmatrix} \quad (5.12)$$

Further a careful readoption of the arguments in section 4.4 will enable us to infer that

(i) $\lim_{n\to\infty} \det\, (\Delta_n(G))^{-1}G \to \partial_0 \neq 0$, a.s.,

(ii) $\lim_{n\to\infty} (\Delta_n(G))^{-1}g \to 0$, a.s., (5.13)

as $n \to \infty$, where $\Delta_n(G)$ is the diagonal matrix with diagonal elements precisely those of G. These observations will imply, through Theorem 2.1

that $\hat{\alpha}$, and hence $\hat{\theta}$ are strongly consistent for the respective parameters.

4.6. TIME SERIES MODELS WITH POLYNOMIAL TRENDS

In this section we shall examine the case when the right side of (1.1) involves a polynomial trend component. We consider the case of a pure trend model for $x(t)$, wherein we let $a_i(t) = t^{i-1}$, $i = 1, \ldots, k$. In order to estimate θ_i the coefficient of t^{i-1} by the method of least squares, the normal equations involve

$$a_{ij} = \sum_{1}^{n} t^{i-1}t^{j-1} = \sum_{1}^{n} t^{i+j-2},\ i, j = 1, 2, \ldots, k. \tag{6.1}$$

Now $a_{ii} \sim (2i-1)^{-1}n^{2i-1} \uparrow \infty$ as $n \to \infty$, and hence

$$\Delta_n^{-1}A'A \sim \begin{bmatrix} 1 & n/2 & \ldots & n^{k-1}/k \\ 3n^{-1}/2 & 1 & \ldots & 3n^{k-2}/k+1 \\ \vdots & & & \\ (2k-1)n^{-k+1}/k & (2k-1)^{-1}n^{-k+2}/k+2 & \ldots & 1 \end{bmatrix} \tag{6.2}$$

We notice that $a_{ij}/a_{ii} \to \infty$ for all $j > i$. This observation will imply, through the Corollary 2.1 that all the components of θ, except the first can be consistently estimated. But, to establish the consistency of $\widehat{\theta_1}$ and hence of $\widehat{\theta}$, we have to evaluate lim $\Delta_n^{-1}A'A$, and establish its non-singularity. We note that

$$|\lim \Delta_n^{-1}A'A| = c \begin{vmatrix} 1 & 1/2 & \ldots & 1/k \\ 1/2 & 1/3 & \ldots & 1/k+1 \\ \vdots & & & \\ 1/k & 1/k+1 & \ldots & 1/2k=1 \end{vmatrix} = c\,|D_k|, \text{ say.} \tag{6.3}$$

The determinant on the right side of (6.3) is non-zero, on making an appeal to the standard result that

$$|D_k| = \{((k-1)!)^4/(2k-1)!(2k-2)!\}|D_{k-1}|. \tag{6.4}$$

Consequently the least squares estimator $\hat{\theta}$ of a pure polynomial trend model is strongly consistent.

4.7 CONCLUDING REMARKS

Theorem 2.1 and its corollary provide sufficient conditions for the strong consistency of least squares estimators of the parameters of a multiple regression model of the type (1.1), when $\{\epsilon(t), \mathcal{F}_t\}$ is a martingale difference sequence.

Applications of these results to specific time series models, discussed in this paper, make it clear that for a non-explosive and trend free time series generated by (1.1) with $a_i(t) = x(t-i)$, $i = 1, \ldots, k$, the least squares estimator of θ is strongly consistent. But, for the purely explosive and partially explosive time series, the least squares method provides consistent estimates only for some specific linear combinations of the parameters. However, in these cases, using auxiliary processes based on the roots of the associated characteristic polynomial, we have established the strong consistency of the least squares estimates of each one of the parameters.

Finally, for a pure polynomial trend model the least squares estimates of the coefficients are seen to be strongly consistent.

REFERENCES

1. Anderson, T. W. (1971): *The Statistical Analysis of Time Series*, John Wiley and Sons Inc. New York.
2. Bhat, B.R. (1982): *Strong consistency of least squares estimates and conditional consistency, Statistics and Probability*, Essays in Honour of C. R. Rao (Ed. Kallianpur, Krishnaiah and Ghosh), North Holland, New York. pp. 105–108.
3. Hall, P. and Heyde, C. C. (1980): *Martingale Limit Theory and its Applications*, Academic Press, New York.
4. Lai, T. L. and Wei, C. Z. (1982): Least square estimator in stochastic regression model with applications to identification and control of dynamic systems, *Annal. Statist.*, **10**, pp. 154–166.
5. ——— (1983): Asymptotic Properties of general autoregressive models and strong consistency of least squares estimates of their parameters, *Jour. of Multi. Anal.*, **13**, pp. 1–23.
6. Venkataraman, K. N. (1968): Some limit theorems on a linear explosive stochastic difference equation with a constant term, and their statistical applications, *Sankhya* **10**, pp. 51–74.
7. ——— (1973): Some Convergence Theorems on a second order linear explosive stochastic difference equation with a constant term, *Jour. Ind. Statist. Assn.*, **11**, pp. 47–69.

5

Efficient Sequential Estimation for Dependent Observations

S. R. Adke*

Miami University, Oxford, Ohio 45056, USA

5.1 INTRODUCTION

The classical theory of point estimation is usually concerned with obtaining a "reasonable" estimator $T(X_1, \ldots, X_n)$, based on a pre-specified, non-random number n of random variables (r.v.'s) $X_1, \ldots, X_n$ for estimating a function $g(\theta)$ of the parameter θ defining their joint distribution. One of the reasonable properties of T is that it be an unbiased estimator. However, it is not always possible to find an unbiased estimator. Consider, for example, the simple situation when $X_1, \ldots, X_n$ are independent and identically distributed (i.i.d.) Bernoulli r.v.'s. with

$$Pr[X_j = 1] = \theta = 1 - Pr[X_j = 0], j = 1, 2, \ldots, 0 < \theta < 1. \quad (1.1)$$

It is well-known that for any fixed n, there is no unbiased estimator of $1/\theta$, $0 < \theta < 1$. [cf. Lehmann (1983), p. 75]. Haldane (1945), who was concerned with this problem of estimating $1/\theta$, suggested that we continue to observe the Bernoulli r.v.'s until a specified number $c > 1$ of them equal unity. The Haldane estimator $(c - 1)/(N - 1)$ is an unbiased estimator of $1/\theta$. Here N, the sample size, is not fixed in advance but is a non-degenerate r.v. The essential features of the Haldane method, which is also called the Inverse Binomial Sampling, to be noted are:

(i) the number N of observations required to construct the estimator is a non-degenerate r.v. which is almost surely finite for every $\theta \in (0, 1)$;

(ii) this r.v. N does not depend on the "future", in the sense that the decision to stop or continue the sampling at any epoch n, depends only on the observations $X_1, \ldots, X_n$ obtained up to that epoch and not on the future observations $X_{n+1}, X_{n+2}, \ldots$.

These two features of the Haldane method constitute the defining features of the method of sequential estimation. Suppose then that we are interested in obtaining a point estimator of a real function $g(\theta)$ of a real

*Present Address: Department of Statistics, University of Poona, Pune 411007, India.

parameter θ occurring in the family of finite dimensional distributions of a real stochastic process $\{X(t), t \in T\}$, where the index set T is either the set Z of non-negative integers or the non-negative half $\mathbb{R}^+ = [0. \infty]$ of the real line $\mathbb{R}$. The extension to multidimensional stochastic processes and to vector-valued parameters is not difficult. In sequential estimation, the period of observation is a r.v. τ taking values in T and it does not depend on the future in the sense that for every $t \in T$, the event $[\tau \leqslant t]$ is $\mathbb{F}_t$-measurable, where $\mathbb{F}_t$ is the σ-algebra $\sigma\{X(u), u \in [0, t] \cap T\}$ generated by the r.v.'s. $X(u), u \in [0, t] \cap T$. Such a r.v. is called a *Markov time* or a *stopping time*. The procedure of sequential estimation is to observe the stochastic process completely over the period $[0, \tau] \cap T$ and use τ and the complete observation $\{X(u), u \in [0, \tau] \cap T\}$ to estimate $g(\theta)$.

In Section 5.2, we assemble a formal apparatus needed to formulate the problem of efficient sequential estimation. The sequential versions of the Cramer-Rao and the Bhattacharyya bounds on the variance of an unbiased estimator of $g(\theta)$ are obtained in a general setting and the problem of efficient sequential estimation is formulated in Section 5.3. Some recent results dealing with efficient estimation are described in Sections 5.4 and 5.5. The last section, Section 5.6, states some open problems.

5.2 THE MODEL AND THE ASSUMPTIONS

Let $\{X(t), t \in T\}$ be a real stochastic process on a probability space $(\Omega, \mathbb{F}, P_\theta)$, where the probability measure P_θ, indexed by $\theta \in \Theta$, is assumed to be complete for each $\theta \in \Theta$. In fact, in the rest of this paper, any statement involving θ, is assumed to hold for all $\theta \in \Theta$, unless otherwise specifically mentioned. The parameter space Θ, is an open, finite or infinite, interval in $\mathbb{R}$.

Let $(\mathbb{R}_T, \mathbb{B}_T)$ be the range space of the process, where $\mathbb{R}_T = \prod_{t \in T} \mathbb{R}_t$ is the Cartesian product of the real lines $\mathbb{R}_t$, and $\mathbb{B}_T$ is the product σ-field $\prod_{t \in T} \mathbb{B}_t$ of the Borel σ-algebras $\mathbb{B}_t$ of $\mathbb{R}_t$, $t \in T$. Let $V(\omega) = \{X(t, \omega), t \in T\}$, $\omega \in \Omega$, be a sample path of the process. The probability measure P_θ^* defined on $\mathbb{B}_T$ by the relation

$$P_\theta^*(F) = P_\theta\{\omega \mid V(\omega) \in F\}, \; F \in \mathbb{B}_T \tag{2.1}$$

is the so called induced probability measure on the sample space $(\mathbb{R}_T, \mathbb{B}_T)$ of the stochastic process. Let $0 < a < b < \infty$ and define $\mathbb{R}_{[a, b]} = \prod_{t \in [a, b]} \mathbb{R}_t$, $\mathbb{B}_{[a, b]} = \prod_{t \in [a, b]} \mathbb{B}_t$ to be the cartesian product space and the product σ-algebra corresponding to closed time interval $[a, b]\ T$, with similar definitions for open or semi-open, semi-closed intervals finite or infinite. The projection $P_{\theta t}$ of P_θ^* on $(\mathbb{R}_{[0, t]}, \mathbb{B}_{[0, t]})$ is defined by

$$P_{\theta t}(F) = P_\theta\{\omega \mid V(\omega) \in F \times \mathbb{R}_{(t, \infty)}\}, \; F \in \mathbb{B}_{[0, t]}, \tag{2.2}$$

and it specifies the probability distribution of the complete observation

$$V(t, \cdot) = \{X(u, \cdot), u \in [0, t] \cap T\}, \tag{2.3}$$

of the process over the set $[0, t] \cap T$, $t \in T$.

Our datum, as stated earlier, is the triplet $[S(\tau), \tau, V(\tau)]$ of the *sampling scheme* $S(\tau)$ consisting of observing the process over $[0, \tau] \cap T$, τ and $V(\tau)$, In order to avoid measure theoretic difficulties, we need the following technical assumption:

A. 1 The stochastic process $\{X(t), t \in T\}$ is a measurable stochastic process, in the sense that the function $X(\cdot, \cdot)$ on $T \times \Omega$ is a measurable function on $(T \times \Omega, \mathbf{B}^T \times \mathbb{F})$ to $(T \times \mathbf{R}, \mathbf{B}^T \times \mathbb{B})$, where $\mathbf{B}^T$ is the σ-algebra of subsets of T.

This assumption ensures that the observation $(\tau, V(\tau))$, defined on $(\Omega, \mathbf{F})$ to $(T \times \mathbf{R}_T, \mathbf{B}^T \times \mathbf{B}_T)$ by the relation $(\tau, V(\tau))(\omega) = \{\tau(\omega), V(\tau(\omega), \omega)\}$ is a measurable function. Let Q_θ be the probability distribution of $\{\tau, V(\tau)\}$:

$$Q_\theta(F) = P_\theta[(\tau, V(\tau)) \in F], F \in \mathbf{B}^T \times \mathbb{B}_T. \tag{2.4}$$

Since our estimation procedure is likelihood based, we assume,

A. 2 There exists a σ-finite measure ν on $(T \times \mathbf{R}_T, \mathbf{B}^T \times \mathbf{B}_T)$, not depending on θ, such that Q_θ is absolutely continuous with respect to ν.

We shall treat the almost everywhere $[\nu]$ unique Radon-Nikodym derivative

$$dQ_\theta/d\nu = L(\tau, V(\tau), \theta) \tag{2.5}$$

as the likelihood of θ corresponding to the observation $(\tau, V(\tau))$. One of the major problems in employing sequential procedures for likelihood based inference is that of obtaining the likelihood function corresponding to a sampling scheme $S(\tau)$. This problem is considerably facilitated for Markov processes by a lemma due to Sudakov (1969), described below. We refer to Kagan, Linnik and Rao (1973, p. 37) for a proof.

Let $\{X(t), t \in T\}$ be a Markov process on $(\Omega, \mathbf{F}, P_\theta)$ with almost surely $[P_\theta]$ right continuous sample paths and suppose assumptions *A*.1 and *A*.2 hold.

Define the Markov time

$$\tau = \inf \{t \mid t \in T) - \{0\}, (t, X(t)) \in S\} \tag{2.6}$$

where S is a closed subset of $T \times \mathbf{R}$. The Markov time τ so defined is in fact the epoch of first entry into set S of the process.

The last or the latest time is said to define a sufficient statistic, if for any $n \geqslant 2$, $t_1, \ldots, t_n \in T$, $0 \leqslant t_1 < \ldots < t_n < \infty$, the statistic $X(t_n)$ is sufficient in the sense that the conditional distribution of $X(t_1), \ldots, X(t_n)$, given $X(t_n)$ does not depend on $\theta \in \Theta$. We assume that for the Markov process $\{X(t), t \in T\}$, the last time defines a sufficient statistic. [cf. Küchler (1982) for a characterization of such Markov processes].

Let $\Lambda_\theta(t)$ denote the probability distribution of $X(t)$ defined by P_θ.

Assume that $\Lambda_\theta(t)$ is absolutely continuous with respect to $\Lambda_{\theta_0}(t)$ for some $\theta_0 \neq \theta$, and suppose that the Radon-Nikodym derivative

$$d\Lambda_\theta(t)/d\Lambda_{\theta_0}(t) = P_{\theta,\theta_0}(t, x)$$

is continuous in its argument $(t, x) \in T \times \mathbf{R}$.

Lemma 2.1 (*Sudakov's Lemma*) Under the assumptions made above about the Markov process $\{X(t), t \in T\}$, the probability measure Q_θ defined by (2.4) corresponding to the first passage Markov time τ defined by (2.6), is absolutely continuous with respect to Q_{θ_0} and

$$dQ_\theta/dQ_{\theta_0} = p_{\theta,\theta_0}(\tau, X(\tau)). \tag{2.7}$$

The Sudakov lemma has been exploited by a number of authors to obtain explicit expressions for $L(\tau, V(\tau), \theta)$, e.g., Winkler and Franz (1979) for the exponential class of multivariate processes with stationary and independent increments, Stefanov (1982a, b) for a particular case of a compound Poisson process, Adke and Manjunath (1984) for continuous time, finite Markov processes, Manjunath (1984a) and (1984b), for ergodic birth-death processes and continuous time, Markov branching processes respectively.

The following lemma due to Stefanov (1982a) is also very useful.

Lemma 2.2 Suppose that $P_{\theta t}$ defined by (2.2) is absolutely continuous with respect to $P_{\theta_0 t}$ for some θ_0 not depending on θ. Let

$$dP_{\theta t}/dP_{\theta t} = L(t, \zeta(t), \theta, \theta_0) \tag{2.8}$$

where $\zeta(t)$ is a measurable function on $[\mathbf{R}_{[0,t]}, \mathbb{B}_{[0,t]}]$ to a measurable space $(Y, \mathbb{B}_Y)$ where Y is a locally compact, separable, metric space and $\mathbb{B}_Y$ is the σ-algebra of its Borel subsets. Let d be a metric on Y. Suppose that $d(\zeta(t_n), \zeta(t)) \to 0$ almost surely $[P_\theta]$ as $t_n \downarrow t$, and that $L(., ., \theta, \theta_0)$ defined by (2.8) is a continuous functions of θ. If τ is an almost surely $[P_\theta]$, finite stopping time, then

$$dQ_\theta/dQ_{\theta_0} = L(\hat{\tau}, \zeta(\hat{\tau}), \theta, \theta_0).$$

5.3. EFFICIENT SEQUENTIAL ESTIMATION

As stated earlier, in a sequential estimation problem, our datum is the triplet $(S(\tau), \tau, V(\tau))$ consisting of the sampling scheme $S(\tau)$, the a.s. finite Markov time τ and the complete observation $V(\tau)$ over the duration $[0, \tau] \cap T$.

Definition 3.1 A sequential plan $S(\tau, g, f)$ is a triplet consisting of the sampling scheme $S(\tau)$ the parametric function $g(\theta)$ and its estimator $f(\tau, V(\tau))$ based on $(\tau, V(\tau))$. The plan is called an unbiased plan if for all $\theta \in \Theta$,

$$E_\theta(f) = \int_{T \times \mathbf{R}_T} f(\tau, V(\tau)) L(\tau, V(\tau), \theta)\, d\nu = g(\theta). \tag{3.1}$$

Introduce the following regularity conditions.

C(i): The unbiased estimator f has finite variance $\sigma^2(f)$ for all $\theta \in \Theta$.

C(ii): The identity

$$\int_{T \times \mathbf{R}_T} L(\tau, V(\tau), \theta)\, d\nu = 1$$

can be differentiated k times, $k \geqslant 1$, under the integral sign.

C(iii): If $W_r = L^{-1}\partial^r L/\partial\theta^r$, $r = 1, \ldots, k$, then the variance-covariance matrix Σ of the random vector $(W_1, \ldots, W_k)'$, where the prime denotes the transpose of the matrix carrying it, is non-singular.

C(iv): The function $g(\theta)$ has finite derivatives

$$g^{(r)}(\theta) = d^r g(\theta)/d\theta^r,\ r = 1, \ldots, k,$$

for all $\theta \in \Theta$, and the relation (3.1) can be differentiated under the integral sign.

Theorem 3.1. Under the assumptions *A*.1, *A*.2, C(i)–C(iv),

$$\sigma^2(f) \geqslant g'(\theta)\Sigma^{-1}g(\theta) = B_k, \text{ say,} \tag{3.2}$$

where $g(\theta) = [g^{(1)}(\theta), \ldots, g^{(k)}(\theta)]'$. Moreover, equality holds in (3.2) iff there exist constants $a_1, \ldots, a_k$, not all zero, such that

$$f(\tau, V(\tau)) - g(\theta) = a_1 W_1 + \ldots + a_k W_k, \tag{3.3}$$

almost surely $[P_\theta]$.

The inequality (3.2) is obtained by observing that the variance covariance matrix of the random vector $(f, W_1, \ldots, W_k)'$ is non-negative definite. The relation (3.2) is a consequence of the fact that the multiple correlation coefficient between f and $(W_1, \ldots, W_k)$ is one, iff f is a linear function of $W_1, \ldots, W_k$.

Observe that B_1 is the sequential version of the Cramer-Rao bound, and that $B_j \leqslant B_{j+1}$ for each $j \geqslant 1$. Various special cases of (3.2) are known in the literature; e.g., when $T = Z^+$, the set of positive integers, and $\{X(t), t \in Z^+\}$ is a sequence of i.i.d. r.v.'s, the inequality (3.2) for $k = 1$ is the Wolfowitz inequality [cf. Wolfowitz (1947)]. Seth (1949) obtained (3.2) for an arbitrary stochastic sequence $\{X(t), t = 1, 2, \ldots\}$. The following definition clarifies what we understand by efficient sequential estimation.

Definition 3.2. An unbiased sequential plan $S(\tau, g, f)$ is *efficient of order* $k \geqslant 2$, or simply *k-efficient* if $\sigma^2(f) = B_k$ for all $\theta \in \Theta$, $\sigma^2(f) > B_j$ for at least one $\theta = \theta_j \in \Theta$, and all $j = 1, \ldots, k-1$. If $\sigma^2(f) = B_1$ for all $\theta \in \Theta$, then $S(\tau, g, f)$ is called 1-efficient. If $S(\tau, g, f)$ is k-efficient for some $k \geqslant 1$, then $S(\tau)$ is called efficient, g is said to be efficient estimable and f is said to be an efficient-estimator.

It should be noted that there do exist plans $S(\tau, g, f)$ for which $\sigma^2(f) \neq B_k$ for all $k \geqslant 1$ and yet $\sigma^2(f)$ does not exceed the variance of any other unbiased estimator $f^*(\tau, V(\tau))$ based on the same Markov time τ and the observation $V(\tau)$. Such an estimator f should be called efficient and the corresponding parametric function g is efficiently estimable. However, in what follows, we restrict ourselves to k-efficient plans only.

The problem of efficient sequential point estimation thus consists of identifying

(i) the sampling scheme $S(\tau)$ which are closed, i.e., for which $P_\theta[\tau < \infty] = 1$ for every $\theta \in \Theta$;

(ii) the class of efficiently estimable parametric functions under the closed sampling schemes;

(iii) the efficient estimators of the efficiently estimable parametric functions; and,

(iv) the uniqueness or otherwise of the efficient estimators.

In the next section, we describe some of the recent results on efficient sequential estimation.

5.4. CRAMER-RAO EFFICIENT ESTIMATION

In this section, we restrict ourselves to 1-efficient estimation which is based on the following sequential version of classical Cramer-Rao inequality:

$$\sigma^2(f) \geqslant [g^{(1)}(\theta)]^2/E\left\{\frac{\partial^2 \log L(\tau, V(\tau), \theta)}{\partial\theta^2}\right\}^2 \tag{4.1}$$

A number of authors have obtained explicit expressions for B_1 for a variety of different stochastic processes and used it to obtain 1-efficient sequential plans.

One of the early results was obtained by DeGroot (1959) who considered a sequence of i.i.d. Bernoulli r.v.'s defined by (1.1). He showed that there are only two efficient sampling schemes, viz., (i) the fixed sample size plan or the simple plan, in which $P_\theta[\tau = n] = 1$ for all $\theta \in \Theta$, and (ii) the inverse sampling plan in which $\tau = \inf\left\{n \,\middle|\, \sum_{i-1}^{n} x_i = c\right\}$ for some fixed positive integer c. Under the simple plan, a linear function of θ and under the inverse plan, a linear function of $1/\theta$, $0 < \theta < 1$ are the only efficiently estimable parametric functions. The results of DeGroot (1959) were generalized to the multinomial populations by Bhat and Kulkarni (1966), whose results, in turn, were extended to finite, discrete parameter irreducible Markov chains by Bai (1975).

Bai considered an irreducible Markov chain $\{X(n), n \in Z\}$ on the state-space $\{1, \ldots, m\}$ with initial distribution

$$\pi : (p_1, \ldots, p_m), p_j > 0, \Sigma\, p_j = 1 \tag{4.2}$$

and transition probabilities

$$p_{ij} = Pr[X_{n+1} = j \mid X_n = i], \quad i, j, \ldots, m$$
$$p_{ij} \geqslant 0, \sum_{j=1}^{m} p_{ij} = 1, \; i = 1, \ldots m, \tag{4.3}$$

which are such that p_{ij} and p_{kj}, $k \neq i$, are functionally independent. Thus the natural parameter space Θ, consists of m-vector $\pi = (p_1, \ldots, p_m)$ and the transition matrix $P = ((P_{ij}))$ satisfying (4.2) and (4.3) and such that the Markov chain is irreducible at all parametric points. Bai demonstrated that if $m > 2$, no function of two or more rows of P is efficiently estimable.

Let $N_{ij}(n)$ denote the number of one-step transitions from state i to state j upto epoch n and let $(\sigma(1), \ldots, \sigma(m))$ be a permutation of $(1, \ldots, m)$. Let $S\left(c, \sum_{j=1}^{k} N_{1\sigma(j)}\right)$ denote the sampling scheme in which $\tau = \inf\{m \mid \Sigma N_{1\sigma(j)}(n) = c\}$, c a positive integer. According to Bai, if one is interested in efficiently estimating a function $g(p_{11}, \ldots, p_{1m})$ without loss of generality, say the first row of P, then under $S(c, \Sigma N_{1\sigma(j)})$ the only efficiency estimable parametric function is

$$c\{\mu_1 p_{11} + \ldots + \mu_m p_{1m}\}/\{p_{1\sigma(1)} + \ldots + p_{1\sigma(k)}\}, \; k \leqslant m,$$

where $\mu_1, \ldots, \mu_m$ are real constants and its efficient estimator is

$$f = \mu_1 N_{11} + \ldots + \mu_m N_{1m}.$$

There are no other efficient sequential plans. In particular, $S\left(c, \sum_{j=1}^{k} N_{ij}\right)$ is efficient and $\sum_{j=1}^{m} \mu_j N_{ij}$ efficiently estimates $\sum_{j=1}^{m} \mu_j p_{ij}$.

Adke and Manjunath (1984) [see also Trybula (1982) and Stefanov (1984)] investigate sequential estimation problems for an irreducible, continuous time, finite Markov process $\{X(t), t \in \mathbb{R}^+\}$ with state-space $\{1, \ldots, m\}$, initial distribution

$$\pi = (p_1, \ldots, p_m)', \; 0 < p_j < 1, \; \Sigma p_j = 1$$

and m-square matrix $Q = ((q_{ij}))$ of intensity rates q_{ij}, $i, j = 1, \ldots, m$, $0 < q_{ij} < \infty$, $i \neq j$, $q_{ii} = -\sum_{k \neq i}^{m} q_{ik}$, $i, j = 1, \ldots, m$. As in the case of the finite Markov chain, one is not able to estimate efficiently the function of two or more rows of Q, when $m > 2$. Suppose then, without loss of generality, that one is interested in estimating a function of the first row of the intensity matrix. Let $N_{ij}(t)$ denote the number of direct transitions from state i to state j during $[0, t]$ and $A_i(t)$ denote the total time spent in state i during $[0, t]$. There are only two types of efficient sequential plans. The plan S_1 employs the stopping time

$$\tau_1 = \inf\left\{t \,\middle|\, \sum_{j=1}^{k} N_{1\sigma(j)}(t) = c\right\}$$

and plan S_2 employs the stopping time,

$$\tau_2 = \inf \{t \mid A_1(t) = t_0\},$$

where c and t_0 are specified positive numbers. Under S_1,

$$c\left\{\mu_1 + \sum_{j=2}^{m} \mu_j q_{ij}\right\} \Big/ \Sigma q_{1\sigma(j)}$$

is efficiently estimated by

$$f_1 = \mu A_1(\tau_1) + \sum_{=2}^{m} \mu_j N_{1j}(\tau).$$

Under S_2, $t_0 \left\{\mu_1 + \sum_{j=2}^{m} \mu q_{ij}\right\}$ is efficiently estimated by

$$f_2 = \mu_1 A_1(\tau_2) + \sum_{j=2}^{m} \mu_j N_{1j}(\tau_2).$$

Winkler and Franz (1979) have dealt with m-dimensional stochastic process $\{X(t) = (X_1(\tau), \ldots, X_m(t))', t \in T\}$ belonging to the exponential class of processes which are processes with stationary and independent increments, for which the probability density function of $X(t)$ with respect to some σ-finite measure ν is

$$\lambda(x, t, \theta) = r(x, t) \exp \{a'(\theta)x + b(\theta)t\}$$

where $x = (x_1, \ldots, x_m)'$, $a(\theta) = (a_1(\theta), \ldots, a_m(\theta))'$, $r(x, t)$ is a non-negative function of x and t only and $a_l(\theta), \ldots, a_m(\theta)$, $b(\theta)$ are non-constant functions of θ only. Note that the dimensions of $X(t)$ and of $a(\theta)$ are the same. This process includes many standard processes like the m-dimensional Wiener process, the multinomial process, the Bernoulli process, the negative binomial process, the Poisson process and the gamma process as particular cases. They exploit the Sudakov Lemma to obtain the likelihood $L(\tau, V(\tau), \theta)$ of θ corresponding to the Markov time τ and the complete observation $V(\tau)$ of the process over $[0, \tau]$, and demonstrate that the exponential class of processes are the only class of processes for which, (i) for every $t \in T$, $X(t)$ is a sufficient statistic to estimate θ, and (ii) for every $\theta_1, \theta_2 \in \Theta$, the probability distributions of $X(t)$ at time t are mutually absolutely continuous and the corresponding Radon-Nikodym derivatives are continuous in (t, x). The Cramer-Rao bound B_1 is obtained and used to specify the class of efficiently estimable parametric functions.

It should be pointed out here that the techniques adopted by all the above mentioned authors are quite similar. One has to obtain: (i) the likelihood $L(\tau, V(\tau), \theta)$; (ii) the Cramer-Rao bound B_1 and necessary and sufficient conditions for its attainability, and (iii) closed sampling schemes and then identify the class of efficiently estimable parametric functions and their efficient estimators.

In obtaining (iii), different authors have employed, (a) the strong laws of large numbers, cf. DeGroot (1959), Bhat and Kulkarni (1966), Bai (1975); (b) completeness of the distribution of the sufficient statistics corresponding to $(\tau, V(\tau))$, cf. Winkler and Franz (1979), Franz (1979), and, (c) the continuous time analogue of the Egorov theorem, cf. Adke and Manjunath (1984), Manjunath (1984a, b).

In all the cases reported above, the parameter space is the natural parametric space. Thus, for example, Trybula (1968), while discussing the sequential plan for a Poisson process, has demonstrated that the simple scheme $P_\theta[\tau = t_0] = 1$, and the inverse scheme $\tau = \inf\{t \mid X(t) = c\}$, c a positive integer, are the only schemes which lead to efficient plans, if one insists on attainment of B_1 for all $\theta > 0$. He then considers a third scheme called the oblique sampling plan for which

$$\tau = \inf\{t \mid X(t) = mt - s\},\ m, s > 0.$$

This stopping time is not closed unless $\theta > m$. However, one can obtain additional efficient sampling plans which are efficient for a restricted parameter space, i.e., which guarantee attainment of B_1 for all $\theta \in \Theta_1, \Theta$, where Θ_1 is a proper subset of Θ. This point of view has been mainly developed by Stafanov in a series of papers, c.f. Stefanov (1982), (1983), (1985a and b). However, such results are applicable only if one knows, a priori, the subset of the parameter space to which θ belongs and this subset satisfies certain very restrictive conditions.

5.5. BHATTACHARYYA EFFICIENT ESTIMATION

It would be noted by virtue of the results reported in section 5.4 that in all cases, the class of efficiently estimable parametric functions, is rather small if one insists on 1-efficient or equivalently on Cramer-Rao-efficient estimators. In an effort to enlarge the class of efficiently estimable parametric functions, Adke and Manjunath (1985) have investigated the class of k-efficient plans by employing the Bhattacharyya bound B_k of order k. They have reported results for the Poisson process, the Bernoulli process, and the two-state, continuous-time Markov process. They demonstrate that the sampling schemes which provide 1-efficient plans, do also lead to k-efficient plans. In particular, for the Poisson process and the Bernoulli process, the simple scheme and the inverse sampling scheme only can be used to k-efficiently estimate polynomials in θ and in $1/\theta$ respectively. In case of the two-state Markov process with intensity matrix

$$\begin{bmatrix} -\lambda & \lambda \\ \mu & -\mu \end{bmatrix},$$

Adke and Manjunath (1985) employ the five schemes proposed by them in their 1984 paper to demonstrate that one can k-efficiently estimate polynomials of degree k in, (i) λ, (ii) μ, (iii) μ^{-1}, (iv) μ^{-1} and a function of the type

$$\Sigma\Sigma\, a_{ij}\lambda^{-i}\mu^{-j}.$$

Although Adke and Manjunath (1985) restrict themselves to the three processes mentioned above, their methods are sufficiently general and can be employed for a number of other processes. In fact, Manjunath (1984b) has provided details for an m-state continuous time Markov process.

These investigations also demonstrate that the class of efficiently estimable parametric functions does not always include the parametric functions of interest. This, in fact, is a severe limitation of the method of minimum variance bound unbiased estimation.

5.6. SOME OPEN PROBLEMS

The following problems are open problems to the best of the author's knowledge and relate to sequential estimation in general and not necessarily only to efficient sequential estimation. An additional list of open problems can be found in Linnik and Romanovsky (1972).

1. Investigate the completeness of the existing plans.
2. Investigate the role of Rao-Blackwellization in sequential estimation.
3. Use other variance bounds like the Chapman-Robbins bound for efficient estimation.
4. Investigate plans based on attainment of a fixed level of Fisher information [cf. Sorenson (1983)].
5. Introduce cost of experimentation in the choice of stopping times.

REFERENCES

1. Adke, S. R. and Manjunath, S. M. (1984), Sequential estimation for continuous time finite Markov processes, *Commun. Statist.—Theor. Meth.*, **13,** 1089–1106.
2. ——— (1985), Efficient sequential estimation for some stochastic processes using Bhattacharyya bounds, Technical Report, **No. 66,** Dept. Statist. Univ. Poona.
3. Bai, D. S. (1975), Efficient estimation of transition probabilities in a Markov Chain, *Ann. Statist.*, **6,** 1305–1317.
4. Bhat, B. R. and Kulkarni, N. V. (1966), On efficient multinomial estimation, *J. Roy. Statist. Soc. Ser. B*, **28,** 45–52.
5. DeGroot, M. H. (1959), Unbiased sequential estimation for binomial population, *Ann. Math. Statist.*, **30,** 80–101.
6. Haldane, J. B. S. (1945), A labour-saving method of sampling, *Nature*, **155,** 49–50.
7. Kagan, A. M. Linnik, Yu. V. and Rao, C. R. (1973), *Characterization Problems in Mathematical Statistics*, J. Wiley and Sons, New York.
8. Kuchler, U. (1982), Exponential families of Markov processes, Part I, General results, Math. Operationfersch. *Statist. Ser. Statist.*, 13, 57–69.
9. Lehmann, E. L. (1983), *Theory of Point Estimation*, J. Wiley and Sons, New York.
10. Linnik, Yu. V. and Romanovski, I. V. (1972), Some new results in sequential

estimation theory, *Proc. Sixth Berkeley Symp. Math. Statist. Prob.*, Ed. LeCam, L. M., Neyman, J., Scott, E. L., **1**, 85–96. Calif. Univ. Press, Berkeley.

11. Manjunath, S. M. (1984a), Optimal sequential estimation for ergodic birth-death processes, *J. Roy. Statist. Soc.*, **B 46**, 412–418.

12. ——— (1984b), Sequential estimation in discrete Markov processes, unpublished Ph.D. thesis, University of Poona, Pune–7, India.

13. Seth, G. R. (1949), On the Variance of Estimates, *Ann. Math. Statist.*, **20**, 1-27.

14. Sorensen, M. (1983), On maximum likelihood estimation in randomly stopped diffusion-type processes, *Internat. Statist. Review*, **51**, 93–110.

15. Stefanov, V. T. (1982b), Sequential estimation for compound Poisson process I, *Serdica*, **8**, 183–189.

16. ——— (1982b), Sequential estimation for compound Poisson Process II, *Serdica*, **8**, 225–261.

17. ——— (1984), Efficient sequential estimation in finite state Markov processes, *Stochastics*, **11**, 291–300.

18. ——— (1985), On efficient stopping times, *Stoch. Processes Their Appl.*, **19**, 305–314.

19. ——— (1985), On the moments of some first passage times, *J. Appl. Prob.*, **22**, 461–466.

20. Sudakov, V. N. (1969), On measures defined by Markovian moments, in *Investigations on Theory of Random Processes*, **Vol. 12** of the Memoirs of the Scientific Seminars of the Leningrad section of the Steklov Institute of Mathematics, Nauka, Moscow, 157–164.

21. Trybula, S. (1968), Sequential estimation in processes with independent increments-*Dissertationes Mathematicae*, **60**, 1–49.

22. ——— (1982), Sequential estimation in finite state Markov processes *Zastosowania Mathematyki*, **17**, 227–248.

23. Winkler, W. and Franz, J. (1979), Sequential estimation problems for the exponential class of processes with independent increments, *Scand J. Statist.*, **6**, 129–139.

24. Wolfowitz, J. (1947), The efficiency of sequential estimates and Wald's equation for sequential processes, *Ann. Math. Statist.*, **18**, 215–230.

6

Outlier Generating Models—A Review

U. Gather

University of Iowa
Technical University of Aachen

and

B. K. Kale*

Iowa State University
University of Poona

6.1 INTRODUCTION

The purpose of this paper is to review some models proposed to explain the occurrence of outliers in a sample $(x_1, \ldots, x_n)$ of real-valued observations. By collecting results from different sources and presenting some new results we investigate the interrelationship between these models and comment upon some merits and demerits of these models.

The problem of statistical analysis of data involving outliers or spurious observations has a long history. The subject originated with the method of Least Squares. Already Legendre (1805) recommended the application of the Method of Least Squares after rejecting those observations whose errors "are found to be such that one judges them too large to be admissible". Bessel (1838) objected to rejecting an observation as inadmissible just because it deviates too much from the majority of observations. The general opinion among scientists using Least Squares in those days is summed up by Edgeworth (1887): "The method of Least Squares is seen to be our best course when we have thrown overboard a certain portion of our data—a sort of sacrifice which has often to be made by those who sail upon the stormy seas of Probability."

Benjamin Peirce (1852) was the first who attempted to define an objective criterion for rejecting outlying observations using probabilistic

---

*Research partially supported by U.S. Army Research Office, Contract No. DAAG-29-85-K-0189.

Both authors wish to thank Professor H. A. David, Iowa State University, for his valuable comments and helpful discussions.

arguments. We quote from Peirce's introduction.

> "In almost every true series of observations, some are found which differ so much from the others as to include some abnormal source of error not contemplated in the theoretical discussions and the introduction of which into the investigations can only serve in the present state of science, to perplex and mislead the investigator. Geometers have, therefore, been in the habit of rejecting those observations which appear to them liable to unusual defects although no exact criterion has been proposed to test and authorize such a procedure. . . ."

He then recommended a principle which first fixed the number, say k, of possible outliers in n observations, and then suggested

> ". . . the proposed observations should be rejected when the probability of the system of errors obtained by retaining them is less than that of the system of errors obtained by their rejection multiplied by the probability of making so many and no more abnormal observations."

Peirce's rule was followed and modified by Chauvenet (1893) whose rule rejected on an average only $\frac{1}{2}$ observation per sample when there is no outlier. Murphy (1951) showed that the level of significance of Chauvenet's test could be as high as 0.4.

Both papers generated considerable amount of controversy. For some criticism and rejoinders we refer to a discussion in Coolidge (1925) as well as in Anscombe (1960). For detailed historical surveys we also refer to Barnett and Lewis (1978, 1984), Grubbs (1950), Ferguson (1961), David (1981), and Hawkins (1980). The modern treatment of outliers can certainly be traced back to the basic ideas of Peirce (1852).

Decision procedures for rejecting outlier observations were developed by many authors, starting with Pearson and Chandrasekar (1936) and later in the works of Murphy (1951), Dixon (1950), Grubbs (1950), Ferguson (1961), and Paulson (1952) among others. They all envisage a situation in which, for a sample of moderate size, one has to decide whether a few observations which are surprisingly far away from the main group of the data are generated by a different random mechanism than anticipated by the experimenter.

On the other hand, the premium-protection approach, initiated by Anscombe (1960), can be viewed as an effort to define estimation procedures for the parameter of interest, avoiding the possible bad influence of outliers.

Indeed, we have come quite far from the viewpoint expressed by Airy (1856) that "the mathematical theory of probabilities fails in all questions applying to errors of extreme magnitude".

This progress is mainly due to the fact that it has been recognized that for an analytical treatment of the outlier problem it is necessary to set up a model alternative to the usual working model or hypotheses adopted by the statistician or experimenter that the observation $(x_1, \ldots, x_n)$ are a random sample from some target distribution $F \in \mathcal{F} = \{F(x, \theta); \theta \in \Omega\}$.

For the problem of testing for outliers, or more explicitly, for the detection of observations from a different distribution, the necessity of an alternative model was felt when the performance of various rejection criteria was studied. For the premium-protection approach such an alternative model is also necessary. Historically, the earliest model is the Mixture-model proposed by Newcomb (1886), Jeffreys (1932), Dixon (1953), and an extensive literature now is available on robust statistical procedures based on mixtures of distributions. However, Tukey (1960) has observed that the mixture model is not appropriate to describe the occurrence of outliers.

Another type of model which is used by many authors implicitly or explicitly is the "k-outlier model", a term coined by David (1979). Here it is assumed that the n observations are still mutually independent, but $(n - k)$ of these are distributed as F, the distribution of the target population, and the remaining k come from a distribution G, but the indexing set of the observations coming from G is not known. For example, such a model for $k = 1$ was used by Anscombe (1960), Anscombe and Barron (1966), Guttman and Smith (1969), Veale and Huntsberger (1969) with F as $N(\mu, \sigma^2)$ and G as $N(\mu + a\sigma, \sigma^2)$ or $N(\mu, b\sigma^2)$. These papers considered the problem of estimating μ and/or σ^2 in the presence of a single observation from a different source. Since all the estimators studied there are symmetric functions of $(X_1, \ldots, X_n)$, it does not matter which particular X_i results from G.

Implicitly using this model, Pearson and Chandrasekar (1936) proposed an outlier test based on $\operatorname*{Max}_i \left|\frac{x_i - \bar{x}}{s}\right|$; see also Grubbs (1950). Dixon (1950) and Murphy (1951) explicitly assumed the above one-outlier model and proposed various tests for outliers.

Ferguson (1961) gave a more general formulation of the k-outlier model. For example, he assumed that $X_1, \ldots X_n$ are mutually independent r.v.'s such that each X_i is $[N(\mu + \sigma\Delta a_{\nu_i}, \sigma^2)]$ where $a_1, \ldots a_n \geqslant 0$ are known constants not all equal and $(\nu_1, \nu_2, \ldots, \nu_n)$ is a permutation of $(1, \ldots, n)$. Thus, the problem of testing for contaminated observations could be viewed as testing for $\Delta = 0$ against $\Delta > 0$. Similar assumptions were made in Gather, Helmers (1983), testing for outliers when the target population is exponential.

Kale [(1969), (1975)] and Kale and Sinha (1971) proposed a modification of the k-outlier model, called exchangeable model, where it was assumed that a priori any set of k r.v.'s out of $(X_1, \ldots, X_n)$ has equal probability $1\Big/\binom{n}{k}$ of being distributed according to G. He showed that under the exchangeable model, the k largest observations are most likely to correspond to the set of r.v.'s distributed as G.

Noting that it is natural to suspect the k-largest observations, Barnett and Lewis (1978) suggested a model which assumes that $X_{(1)}, X_{(2)}, \ldots, X_{(n-k)}$, the first k components of the order statistics correspond to r.v.'s distributed as F and the last k-components $X_{(n-k+1)}, \ldots, X_{(n)}$ correspond to G. It may be emphasized here that this model, called labelled alternative, specifies the joint distribution of the order statistics of the sample rather than the joint distribution of $(X, \ldots, X_n)$ specified by the k-outlier model or the corresponding exchangeable model.

The object of this paper is to discuss the above distribution models proposed to explain the occurrence of outliers. In the first part we give formal descriptions of these models, study some distributional properties and shortly sketch Maximum Likelihood Estimation in the forementioned models. However, for specific inference procedures we cannot do better than refer to Barnett and Lewis (1984) where these are studied in full detail. We also confine only to the above models, and do not include the deterministic or so-called identified outlier model where the indices of the contaminated observations are supposed to be known.

So far we have considered the situation in which outliers are caused by a different distribution G. On the other hand one can envisage a situation where all observations come from some possible "outlier prone" distribution H causing a few observations to be removed far away from the main group of data. Neyman and Scott (1971) and later Green (1976) presented and analyzed the concept of outlier proneness and outlier resistance of distributions. In section 6.3 we will examine the above mentioned outlier generating models from this viewpoint and discuss the statistical implications of these concepts. In section 6.5 we will briefly discuss the Bayesian approach and present some concluding remarks and comments.

6.2. OUTLIER-GENERATING MODELS

6.2.1. The k-outlier Model

As mentioned in the introduction, in the k-outlier model we assume that our observations $x_1, \ldots, x_n$ result from independent random variables $X_1, \ldots, X_n$, $(n - k)$ of which are distributed as $F \in \mathcal{F}$ while the remaining k are distributed as $G \in \mathcal{G}$. Here, $\mathcal{F}$ and $\mathcal{G}$ are (parametric) classes of distributions and we allow $\mathcal{G} = \mathcal{G}(F)$ to depend on F. The indices of the observations from G are supposed to be unknown. We will denote this model by M_k and consider first the case $k = 1$. We establish some useful results for M_1 and indicate how these can be generalized to M_k for $k > 1$.

Denote by f and g the p.d.f.'s of F and G respectively and consider the unknown index ν of the random variable which is distributed according to G as parameter with parameter space $N = \{1, 2, \ldots, n\}$. Thus the joint p.d.f. of $(X_1, \ldots, X_n)$ under model M_1 is given by

$$L(\mathbf{x} \mid f, g, \nu) = g(x_\nu) \prod_{i \neq \nu} f(x_i),\ \nu \in N,\ F \in \mathcal{F},\ G \in \mathcal{G}. \qquad (2.1.1)$$

For example, if f and g involve location and scale parameters, we have

$$L(\mathbf{x} \mid \mu, \sigma, \delta, \nu) = \frac{1}{\sigma} f\left(\frac{x_\nu - \mu - \delta\sigma}{\sigma}\right) \prod_{i \neq \nu} \frac{1}{\sigma} f\left(\frac{x_i - \mu}{\sigma}\right), \qquad (2.1.2)$$
$$(\mu, \sigma, \delta) \in \Theta,\ \nu \in N$$

which represents the case of shift in location for the outlier-observation from μ to $\mu + \delta\sigma$. Similarly

$$L(\mathbf{x} \mid \mu, \sigma, \lambda, \nu) = \frac{1}{\lambda\sigma} f\left(\frac{x_\nu - \mu}{\lambda\sigma}\right) \prod_{i \neq \nu} \frac{1}{\sigma} f\left(\frac{x_\nu - \mu}{\sigma}\right), \qquad (2.1.3)$$
$$(\mu, \sigma, \lambda) \in \Theta,\ \nu \in N$$

represents the case where the scale parameter has changed. Θ in both cases denotes an appropriate parameter-space.

It is important to note that for $G = F$ in (2.1.1), that is for $\delta = 0$ or $\lambda = 1$ in (2.1.2) or (2.1.3) respectively, we get the homogeneous model where $X_1, \ldots, X_n$ are i.i.d. random variables; this we will denote by H_0.

A model as in (2.1.1) was indeed assumed implicitly or explicitly by most of the workers in the field. Deriving performance-characteristics of inference procedures, they often assumed that $\nu = n$, which effectively meant that X_n is claimed to be the source of heterogeneity, being distributed as G. This assumption is sometimes referred to as 'identified outlier model', explicitly used for example by Veale (1975). But as far as inference procedures are considered which are based on the order statistics $X_{(1)} \leqslant X_{(2)} \leqslant \ldots \leqslant X_{(n)}$, the performance characteristic evaluated by assuming $\nu = n$ remain valid for M_1. We state this as a general result, which is obtained by an obvious substitution.

Theorem 2.1.1. If $T(x_1, \ldots, x_n)$ is a symmetric statistic, then in the model M_1, the distribution of T is independent of $\nu \in N$.

In particular, for $T(x_1, \ldots, x_n) = (x_{(1)}, \ldots, x_{(n)})$ being the order statistics, this result was noted by Veale (1972), and for some other outlier-generating models, including M_1, also by Gather (1984) and it has been used implicitly before by many workers in this area.

Because of the above theorem, the joint distribution of the order statistics $(X_{(1)}, \ldots, X_{(n)})$ does not depend upon $\nu \in N$, which implies that the order statistics cannot be sufficient for the class M_1 of distributions, as they all depend on ν.

In fact one can show that for M_1, in general, there does not exist any non-trivial sufficient statistic. We give the proof here for the case where f and g both belong to a common one-parameter exponential family of densities with parameters θ_1 and θ_2 respectively. The joint p.d.f. then is given by

$$L(\mathbf{x} \mid \theta_1, \theta_2, \nu) = \beta(\theta_2) \exp \{\theta_2 T(x_\nu)\}\{[\beta(\theta_1)]^{n-1} \exp \{\theta_1 \sum_{i \neq \nu} T(x_i)\}$$

$$\times \prod_{i=1}^{n} h(x_i),$$

where $\theta := (\theta_1, \theta_2) \in \Omega$, where Ω is the natural range of the parameter space.

If then $\mathbf{x}$ and $\mathbf{y}$ are both points in the support of L which belong to the same partition, induced by some sufficient statistic, the ratio $\dfrac{L(\mathbf{x} \mid \theta_1, \theta_2, \nu)}{L(\mathbf{y} \mid \theta_1, \theta_2, \nu)}$ and thus $\theta_1[\sum_{i \neq \nu} T(x_i) - \sum_{i \neq \nu} T(y_i)] + \theta_2[T(x_\nu) - T(y_\nu)]$ does not depend on $(\theta_1, \theta_2, \nu)$. This implies that $T(x_\nu) = T(y_\nu)$ for all $\nu = 1, 2, \ldots, n$. This simple argument extends to the multiparameter exponential family with p.d.f.

$$f(x, \theta_1, \ldots, \theta_m) = \beta(\boldsymbol{\theta}) \exp \left\{ \sum_{i=1}^{m} \theta_i T_i(x) \right\} h(x) \text{ and } g = f(x, \boldsymbol{\theta}'), \boldsymbol{\theta}' \neq \boldsymbol{\theta}.$$

Noting that the order statistics together with the rank order statistics $(R_1, \ldots, R_n)$, where $R_i = \text{Rank}(X_i)$, are equivalent to the sample $(X_1, \ldots, X_n)$ itself, it follows that all inference in M_1 can be based on those two together.

However, if we are not interested in the parameter ν but only in the parameters occurring in f and/or g, then we can restrict to procedures based on the order statistics or other symmetric functions of the observations. Thus the use of order statistics is justified when estimating the parameters μ and/or σ and even δ or λ occurring in (2.1.2) or (2.1.3), and also when testing for the existence of an outlier, that is testing for $\delta > 0$, in (2.1.2) for instance, as was done by all the researchers in the past. In fact, the joint distribution of $(X_{(1)}, \ldots, X_{(n)})$ can be shown to be the partial likelihood, as defined by Cox (1975), or conditional likelihood as defined by Godambe (1976), noting that the marginal distribution of $(R_1, \ldots, R_n)$ depends on (θ, ν), but the conditional distribution of the order statistics given $(R_1, \ldots, R_n)$ does not depend on ν.

Another reason to confine ourselves to symmetric statistics arises if the decision problem we are considering is invariant with respect to the group of symmetric transformations, as was done by Murphy (1951) and Ferguson (1961) for the problem of detecting the existence of an outlier without identifying the same, and also for inference about f and/or g only.

If we have to identify the outlier, that is ν, then we must use the rank order statistics also. This is precisely the current practice; e.g. for $F \sim N(\mu, \sigma^2)$ and $G \sim N(\mu + \delta\sigma, \sigma^2)$, with $\delta > 0$, already Pearson and Chandrasekar (1936) recommended using $\tau_{(n)} := (x_{(n)} - \bar{x})/s$ and rejecting H_0, the null-hypothesis of no outliers, if $\tau_{(n)} > \tau_0$, and identifying $x_{(n)}$ as an outlier in that case, and proceeding analogously when $\delta < 0$, using $\tau_{(1)} := (x_{(1)} - \bar{x})/s$, and $\max(\tau_{(n)}, -\tau_{(1)})$ when the sign of δ is unknown.

Viewing the identification of an outlier as a slippage-problem in M_1, Kudo (1956) and later Hall and Kudo (1968) showed the following optimality property of the above mentioned procedure. In a multidecision set-up, the above procedure yields the maximal probability of identifying the crucial index ν correctly, under all symmetric, location- and scale-invariant multi-decision procedures with given Type I error-probability of rejecting H_0 incorrectly.

Traditionally, only extreme observations have been suspected as outliers and we show below that under certain restrictions on F and G, this is a natural thing to do. Although these results were proved for the exchangeable model $M_1(E)$, which will be considered in the next section, they are valid for M_1 as well.

If, $R_i := \text{Rank}(X_i)$ as above, we will consider the following probabilities

$$u_r := P(R_\nu = r) \tag{2.1.4}$$

$$= \binom{n-1}{r-1} \int_{-\infty}^{\infty} [F(y)]^{r-1}[1 - F(y)]^{n-r}\, dG(y),\ r = 1, 2, \ldots, n$$

which is seen by a simple combinatorial argument.

Note that (2.1.4) does not depend on ν. Using integral transformation, following Mount and Kale (1973), we have

$$u_r = \binom{n-1}{r-1} \int_0^1 y^{r-1}(1-y)^{n-r}\Psi[F^{-1}(y)]\, dy \tag{2.1.5}$$

where $\Psi(x) := \dfrac{\partial G(x)}{\partial F(x)} = \dfrac{g(x)}{f(x)}$. If we assume that Ψ is monotone in x, then using the MLR property of the beta distribution in r, it follows immediately that u_r is monotone in r. Thus u_n (or u_1) is the maximum; that is, roughly speaking, the observation from G (the outlier) is most likely to be the observation with largest (or smallest) rank, thus $x_{(n)}$ (or $x_{(1)}$).

In case that $G(x) = F(x - \delta)$ or $G(x) = F\left(\dfrac{x}{\lambda}\right)$ as in (2.1.2) or (2.1.3), the probabilities $u_r = u_r(\delta)$, or $u_r = u_r(\lambda)$ respectively, have nice asymptotic properties in the limiting cases $\delta \to \infty$, $(\delta \to -\infty)$ or $\lambda \to \infty$, $(\lambda \to 0)$. As is expected, using the Lebesgue Dominated Convergence Theorem, we have

$$\lim_{\delta\to\infty} u_r(\delta) = \begin{cases} 0, \text{ if } r < n \\ 1, \text{ if } r = n \end{cases};\ \lim_{\delta\to-\infty} u_r(\delta) = \begin{cases} 0, \text{ if } r > 1 \\ 1, \text{ if } r = 1 \end{cases} \tag{2.1.6a}$$

when $G(x) = F(x - \delta)$ and when $G(x) = F\left(\dfrac{x}{\lambda}\right)$,

$$\lim_{\lambda\to\infty} u_r(\lambda) = \begin{cases} 0, & 1 < r < n \\ F(0), & r = 1 \\ 1 - F(0), & r = n \end{cases};\ \lim_{\lambda\to 0} u_r(\lambda) = \binom{n-1}{r-1}[F(0)]^{r-1}(1 - F(0))^{n-1} \tag{2.1.6b}$$

From (2.1.6a), we deduce that, with increasing shift of location of the contaminated distribution G, the largest (or smallest) observation is with high probability, the one that results from G. Analogously, (2.1.6b) shows that, quite naturally, the probability of $x_{(n)}$ or $x_{(1)}$ being the observation from the distribution with scale tending to infinity, depends on $F(0)$. It also shows that we have to expect the observation from G in the central part of the sample if the scale of G tends to zero.

Other asymptotic results in M_1 are those by David and Shu (1978) and David (1985) about the limit behaviour of the expected values and variance-covariances of components of order statistics in (2.1.2) and (2.1.3) as $\delta \to \pm\infty$ or $\lambda \to \infty$ respectively. In general, one can show that if $\delta \to \infty$ ($\delta \to -\infty$), then $X_{(1)}, \ldots, X_{(n-1)}$ (or $X_{(2)}, \ldots, X_{(n)}$) behave like the order statistics of a sample of size $n-1$ from F and also that as $\lambda \to \infty$, $X_{(2)}, \ldots, X_{(n-1)}$ would behave as order statistics of a sample of size $n-2$ from F. The special status of the extreme order statistics in model M_1 becomes also obvious when we consider Maximum-Likelihood estimation of the parameters in this model.

We assume that $\Psi(x) = \dfrac{\partial G(x)}{\partial F(x)}$ is strictly increasing and illustrate our results once more for the special case of a one parameter exponential class of densities, where

$$L(\mathbf{x} \mid \theta_1, \theta_2, \nu) = [\beta(\theta_1)]^{n-1}\beta(\theta_2) \exp \{\theta_1 \sum_{i \neq \nu} T(x_i) + \theta_2 T(x_\nu)\} \prod_{i=1}^{n} h(x_i)$$

$$= [\beta(\theta_1)]^{n-1}\beta(\theta_2) \exp \left\{\theta_1 \sum_{i=1}^{n} T(x_i) + (\theta_2 - \theta_1) T(x_\nu)\right\} \pi h(x_i) \tag{2.1.7}$$

Assuming that $\theta_2 > \theta_1$ and $T(x)$ is strictly increasing (or $\theta_2 < \theta_1$ and $T(x)$ is strictly decreasing) so that $\Psi(x) = \dfrac{\partial G}{\partial F}$ is strictly increasing in x, it follows that (2.1.7) is maximized w.r.t. ν for each $(\theta_1, \theta_2) \in \Omega$ by $\hat{\nu}$ defined by

$$\text{Rank}\,(X_{\hat{\nu}}) = R_{\hat{\nu}} = n \tag{2.1.8}$$

Now maximizing (2.1.7) w.r.t. (θ_1, θ_2) with $\nu = \hat{\nu}$ fixed, we obtain

$$\hat{\theta}_1 = \rho^{-1}\left[\frac{1}{n-1} \sum_{i=1}^{n-1} T(X_{(i)})\right] \text{ and } \hat{\theta}_2 = \rho^{-1}[T(X_{(n)})]. \tag{2.1.9}$$

where $\quad \rho(\theta) = \dfrac{-\partial \log \beta(\theta)}{\partial \theta}.$

For the special case of the exponential distribution with $\theta_2 = \lambda\theta_1$ Gather (1985) obtained these estimates. See also Kale (1975). The maximum likelihood estimation in the multiparameter case (even if restricted to an exponential class of densities) is much more complex. For the special case $F \sim N(\mu_1, \sigma^2)$ and $G \sim N(\mu_2, \sigma^2)$ we can show that if $\mu_2 > \mu_1$

then $\hat{\nu}$ is given by (2.1.8) and

$$\hat{\mu}_1 = \frac{1}{n-1} \sum_{i=1}^{n-1} X_{(i)} \quad \hat{\mu}_2 = X_{(n)}, \quad \hat{\sigma}^2 = \frac{1}{n} \sum_{i=1}^{n-1} (X_{(i)} - \hat{\mu}_1)^2.$$

For the case of change in scale only where $F \sim N(\mu, \sigma_1^2)$ and $G \sim N(\mu, \sigma_2^2)$ with $\sigma_2^2 > \sigma_1^2$, we refer to Small and McLeish (1985). It would be worthwhile to investigate in more detail and generality the maximum likelihood estimation in M_1 when F and G have densities belonging to a multiparameter exponential class.

As a straightforward generalization of M_1, in the k-outlier model M_k, the joint p.d.f. of $X_1, \ldots, X_n$ with $F \in \mathcal{F}$ and $G \in \mathcal{G}$, having densities f and g respectively, is given by

$$L_k(\mathbf{x} \mid f, g, \nu) = \{ \prod_{i \in \nu} g(x_i)\} \prod_{i \notin \nu} f(x_i), \; \nu \in N_{(k)}, \; F \in \mathcal{F}, \; G \in \mathcal{G} \quad (2.1.10)$$

Thus ν, the indexing subset of observations coming from G is an element of $N_{(k)}$, the collection of all $\binom{n}{k}$ subsets of selecting k integers out of n. The basic result, given in Theorem 2.1.1, remains valid here also, that is, the distribution of a symmetric function of $(X_1, \ldots, X_n)$ does not depend on $\nu \in N_{(k)}$. Also, the restriction to inference procedures based on $(X_{(1)}, \ldots, X_{(n)})$ is justifiable for the same reasons as in M_1. For $\nu = \{j_1, \ldots, j_k\} \in N_{(k)}$, we denote $(\gamma_1, \ldots, \gamma_k) = (R_{j_1}, \ldots, R_{j_k}) =$ [Rank $(X_{j_1}), \ldots,$ Rank (X_{j_k})]. Using arguments similar to those given in Kale (1974), we get that the distribution of $(\gamma_1, \ldots, \gamma_k)$ does not depend on ν. Further, the results in Kale (1975) for the exchangeable model $M_k(E)$ are valid here too, that is, it holds that the probability $\phi(r_1, \ldots, r_k) = P(\gamma_1 = r_1, \ldots, \gamma_k = r_k)$ is monotone in each of its arguments and $\max_{N_{(k)}} \phi(r_1, \ldots, r_k) = \phi(n-k+1, \ldots, n)$ provided $\frac{\partial G(x)}{\partial F(x)}$ is monotone increasing. This means that the k largest observations $x_{(n-k+1)}, \ldots, x_{(n)}$ are most likely to be corresponding to those observations distributed as G, or the MLE of ν is given by

$$\{R_{\hat{j}_1}, \ldots, R_{\hat{j}_k}\} = \{(n-k+1), \ldots, n\}$$

As in the case $k = 1$, this result can be used in M_k to justify suspecting the k largest (smallest) observations as outliers and we can show that for the shift in location case where $G(x) = F(x - \delta)$

$$\lim_{\delta \to \infty} \phi(r_1, \ldots, r_k) = \begin{cases} 1, & r_1 = n-k+1, \ldots, r_k = n \\ 0, & \text{otherwise} \end{cases}$$

Similar results are valid for $\delta \to -\infty$, and $\lambda \to \infty$ when $F(0) = 0$, while we have a more complex behaviour for $\lambda \to \infty$ when $F(0) \neq 0$.

When $k > 1$, it might be the case that not all outliers have the same source of heterogeneity, but are produced by $r > 1$ different distributions $G_1, \ldots, G_r$ with multiplicities $n_1, \ldots, n_r$ where $\sum_{i=1}^{r} n_i = k$. We denote $(F_{\nu_1}, \ldots, F_{\nu_n})$, the d.f.'s of the mutually independent r.v.'s $X_1, \ldots, X_n$, where $(\nu_1, \ldots, \nu_n) =: \nu \in \pi_n$ now is an unknown permutation of $(1, 2, \ldots, n)$ and $F_1 = F_2 = \ldots = F_{n-k} = F$ and $F_{n-k+i} = G_i$, with densities $f, g_i, i = 1, 2, \ldots, k$, with G_i's possibly all different or all the same, or any combination in between. The joint p.d.f. of $(X_1, \ldots, X_n)$ in this model, which we will denote by $M_{n_1, \ldots, n_r}$, is given by

$$L(\mathbf{x} \mid f, g_1, \ldots, g_r, \nu) = \prod_{i=1}^{n} f_{\nu_i}(x_i),\ \nu = (\nu_1, \ldots, \nu_n) \in \pi_n, \quad (2.1.11)$$

$$F_i = F \in \mathcal{F},\ i = 1, \ldots, n-k$$

$$F_j = G_j \in \mathcal{G},\ j = n-k+1, \ldots, n$$

As before restriction to the order statistics is justified if we are interested in parameters occurring in f and/or g_i, $i = 1, \ldots, k$ while ignoring the parameter ν. Restriction to symmetric functions of observations can also be justified when considering decision problems which are invariant w.r.t. this group of transformations over $M_{n_1, n_2, \ldots, n_r}$. Ferguson (1961) taking $F \sim N(\mu, \sigma^2)$ and $F_j \sim N(\mu + \delta a_j \sigma, \sigma^2)$ with known a_j, such that $\Sigma (a_j - \bar{a})^2 > 0$ derived an LMP invariant test for the existence of k upper outliers. This test is based on b_1, the sample skewness. On the other hand for change in scale only, Ferguson showed that an LMP invariant test for k outliers is based on b_2, the sample kurtosis. Similar investigations were carried out by Gather and Helmers (1983) for the exponential distribution for the case of change in scale only, and for more general shift in location or change in scale problems in Gather (1984).

However, even though a few outliers may come from different sources, this further subdivision among the r distributions generating k outliers increases the number of parameters very heavily and requires knowledge or assumptions about the relation between these parameters and/or distributions. This may not often be possible as well as advisable since in the k-outlier model one assumes k to be small in relation to n.

6.2.2. The Exchangeable Model

The exchangeable model for one outlier, which we will denote by $M_1(E)$, was introduced by Kale (1969) and applied to the exponential distribution with location zero and scale σ by Kale and Sinha (1971). For the general k-outlier situation, $M_k(E)$ was introduced by Kale (1975).

It is envisaged here that prior to the experiment, any set of k r.v.'s out of $(X_1, \ldots, X_n)$ has equal probability $1\Big/\binom{n}{k}$ of corresponding to the nonhomogenous distribution G, while the other r.v.'s are distributed according to F, whereas in M_1, we assume $F \in \mathcal{F}$, $G \in \mathcal{G}$ and denote the densities of F and G by f and g respectively. More precisely, the parameter $\nu \in N_{(k)}$ is assumed to have a uniform prior over the parameter space $N_{(k)}$ and the conditional joint density of $(X_1, \ldots, X_n)$, given ν, is equal to $L(\mathbf{x} \mid f, g, \nu)$ from (2.1.1). Hence, by summing over the parameter ν we get for the joint p.d.f. of $(X_1, \ldots, X_n)$ in the special case $M_1(E)$

$$L(\mathbf{x} \mid f, g) = \frac{1}{n} \sum_{\nu=1}^{n} g(x_\nu) \cdot \{ \prod_{i \neq \nu} f(x_i) \}, \; F \in \mathcal{F}, G \in \mathcal{G} \tag{2.2.1}$$

and in case of $M_k(E)$, we have

$$L_k(\mathbf{x} \mid f, g) = \frac{1}{\binom{n}{k}} \sum_{\nu \in N_{(k)}} \{ \prod_{j \in \nu} g(x_j) \} \cdot \{ \prod_{i \notin \nu} f(x_i) \}, \; F \in \mathcal{F}, G \in \mathcal{G} \tag{2.2.2}$$

It is important to note that if $T(x_1, \ldots, x_n)$ is a symmetric function of the observations, then the distribution of T under M_k and $M_k(E)$ is the same, as is proven in Gather (1984), and follows from an obvious substitution.

In model M_k, restriction to symmetric functions of $(X_1, \ldots, X_n)$ was justified on the basis of conditional likelihood and/or invariance principles, though the order statistic $(X_{(1)}, \ldots, X_{(n)})$ was not sufficient for M_k. On the other hand, since under (2.1.2) the r.v.'s $X_1, \ldots, X_n$ are exchangeable, though dependent, the order statistic $(X_{(1)}, \ldots, X_{(n)})$ is sufficient for $M_k(E)$, which makes it natural to restrict oneself to inference procedures based on the order statistics.

The probability u_r, derived in (2.1.4) for M_1, now can be interpreted as the posterior probability of $X_{(r)}$ resulting from G. Similarly the probability $\phi(r_1, \ldots, r_k)$ can be viewed as the posterior probability of the event that $X_{(r_1)}, \ldots, X_{(r_k)}$ come from G. From the results of Kale (1975), $\phi(r_1, \ldots, r_k)$ is strictly increasing, as in (2.1.5), and we have the same asymptotic behaviour as in (2.1.6) of $\phi_\delta(r_1, \ldots, r_k)$ or $\phi_\lambda(r_1, \ldots, r_k)$ say, for $\delta \to \pm\infty$ or $\lambda \to \infty$ in the shift in location case where $G(x) = F(x - \delta)$ or the change in scale case with $G(x) = F(x/\lambda)$ respectively.

Also, for $G(x) = F(x - \delta)$, in the limit, $(X_{(1)}, \ldots, X_{(n-k)})$ behave like the order statistic of a sample of size $(n-k)$ from F and $(X_{(n-k+1)}, \ldots, X_{(n)})$ as the order statistics of a sample of size k from G.

More generally, Bendre (1985) has shown that for any sequence $\{G_\lambda\}$ with $\frac{\partial G_\lambda(x)}{\partial F(x)}$ strictly increasing in x for all λ, and $\lim_{\lambda \to \infty} G_\lambda(a) = 0$ for all $a \in R$, for large λ, the joint p.d.f. of $(X_{(1)}, \ldots, X_{(n)})$ is approximately equal to

$$h(x_{(1)}, \ldots, x_{(n)}) = \frac{(n-k)!\, k!}{\phi(n-k+1, \ldots, n)} \prod_{i=1}^{n-k} f(x_{(i)}) \prod_{i=n-k+1}^{n} g_\lambda(x_{(i)}) \tag{2.2.3}$$

This property will lead us most naturally to the labelled slippage alternatives, first proposed by Barnett and Lewis (1978) which will be studied in some detail in the next section.

We consider now, as we did for M_1, Maximum-Likelihood estimation of the parameters in $M_1(E)$, where F and G belong to one parameter exponential families with parameters θ_1 and θ_2 respectively, such that we have as joint p.d.f. of $(X_1, \ldots, X_n)$,

$$L_1(\mathbf{x} \mid \theta_1, \theta_2) = \frac{1}{n} \sum_{\nu=1}^{n} \beta(\theta_2)[\beta(\theta_1)]^{n-1} \exp\{\theta_2 T(x_\nu) + \theta_1 \sum_{i \neq \nu} T(x_i)\} \prod_{i=1}^{n} h(x_i).$$

The likelihood equations for (θ_1, θ_2) are given by

$$\sum_{\nu=1}^{n} \left[(n-1)\frac{\partial \beta(\theta_1)}{\partial \theta_1} + \beta(\theta_1) \sum_{i \neq \nu} T(x_i)\right] \exp\{(\theta_2 - \theta_1)T(x_\nu)\} = 0$$

$$\sum_{\nu=1}^{n} \left[\frac{\partial \beta(\theta_2)}{\partial \theta_2} + \beta(\theta_2)T(x_\nu)\right] \exp\{(\theta_2 - \theta_1)T(x_\nu)\} = 0 \tag{2.2.4}$$

which shows that the MLEs of (θ_1, θ_2) in M_1 and $M_1(E)$ are different, which is mainly due to the fact that in $M_1(E)$, the parameter ν is integrated out using a uniform prior whereas in M_1 it was estimated. Result (2.2.4) also shows that the trimmed means are not connected with MLEs in $M_k(E)$ as was claimed in Kale (1975). In fact, as we have seen in Section 6.2, the estimators given by Kale (1975) are MLEs in M_1 rather than in $M_1(E)$. There is a considerable amount of literature on estimation of location of the normal distribution and the scale of the exponential distribution based on the premium protection approach of Anscombe (1960). For references we refer to Barnett and Lewis (1984) and Gather (1986a), and Rauhut (1982).

The case, where the outliers might bave different sources of heterogeneity, as was discussed in (2.1.9) for M_1, has a straightforward analogue $M_{n_1, \ldots, n_r}(E)$, which we do not study any further. However, it is worthwhile pointing out that the locally most powerful tests for the existence of k outliers from location-shift-, or scale-inflated distributions have the same optimality-properties in the exchangeable models $M_{n_1, \ldots, n_r}(E)$, since these properties depend only on the distribution of the order statistics.

6.2.3 The Labelled Model

As pointed out earlier, it is natural to suspect the largest observations as outliers. This idea is incorporated in the labelled slippage alternatives, introduced by Barnett and Lewis (1978). It may be pointed out at the outset that as opposed to $M_k(E)$ and M_k, the labelled model $M_k(L)$ is specified in terms of the distribution of the order statistic assuming that the largest (smallest) k observations come from G and the remaining ones from F.

More precisely, one assumes that the order statistic $(X_{(1)}, \ldots, X_{(n)})$ is distributed as $(Y_{(1)}, \ldots, Y_{(n-k)}, V_{(1)}, \ldots, V_{(k)})$ where $Y_1, \ldots, Y_{n-k}, V_1, \ldots, V_k$ all are mutually independent and $Y_1, \ldots, Y_{n-k}$ are distributed according to F, and $V_1, \ldots, V_k$ according to G, and further

$$\text{Max}\,(Y_1, \ldots, Y_{n-k}) \leqslant \text{Min}\,(V_1, V_2, \ldots, V_k).$$

Hence the joint p.d.f. of $(X_{(1)}, \ldots, X_{(n)})$ is given by

$$h(x_{(1)}, \ldots, x_{(n)} \mid f, g) = \frac{(n-k)!\, k!}{\phi_k(F, G)} \left\{\prod_{i=1}^{n-k} f(x_{(i)})\right\}\left\{\prod_{i=n-k+1}^{n} g(x_{(i)})\right\},$$
$$x_{(1)} \leqslant \ldots \leqslant x_{(n)}$$
$$F \in \mathcal{F}, G \in \mathcal{G}, \qquad (2.3.1)$$

where $\phi_k(F, G)$ denotes the probability of the event $\{\text{Max}\,(Y_1, \ldots, Y_{n-k}) \leqslant \text{Min}\,(V_1, \ldots, V_k)\}$. Note that (2.3.1) has the structure of the asymptotic density in (2.2.3) in the exchangeable model.

We note that the joint distribution of the order statistics $(X_{(1)}, \ldots, X_{(n)})$ in $M_1(L)$ can be derived from the distribution of $(X_{(1)}, \ldots, X_{(n)})$ under M_1 or $M_1(E)$ by restricting oneself to the term corresponding to u_n and ignoring the remaining $(n-1)$ terms. Thus in the location shift case since $u_n(\delta) \to 1$ as $\delta \to \infty$, for large δ the results for M_1, $M_1(E)$ and $M_1(L)$ would be very similar. However, the rate of convergence of $u_n(\delta)$ to one may be slow and extremely large values of δ may be required to approximate the p.d.f. under M_1 or $M_1(E)$ by that under $M_1(L)$ with a desired level of accuracy. This aspect needs further detailed investigation.

Another important distinction in $M_1(L)$ and M_1 or $M_1(E)$ becomes clear when we consider MLE's of (θ_1, θ_2) in $M_1(L)$ where f and g belong, once more, to a one parameter exponential family. Then the joint p.d.f. of the order statistics is,

$$L_1(x_{(i)} \mid \theta_1, \theta_2) = \frac{(n-1)!}{\phi_1(\theta_1, \theta_2)} [\beta(\theta_1)]^{n-1}\beta(\theta_2) \exp\left\{\theta_1 \sum_{i=1}^{n-1} T(x_{(i)}) + \theta_2 T(x_{(n)})\right\}$$
$$\times \prod_{i=1}^{n} h(x_{(i)})$$

and the likelihood equations are

$$\sum_{i=1}^{n-1} T(x_{(i)}) = \frac{\partial \log \phi_1(\theta_1, \theta_2)}{\partial \theta_1} - (n-1)\frac{\partial \log \beta(\theta_1)}{\partial \theta_1}$$
$$T(x_{(n)}) = \frac{\partial \log \phi_1(\theta_1, \theta_2)}{\partial \theta_2} - \frac{\partial \log \beta(\theta_2)}{\partial \theta_2} \qquad (2.3.2)$$

Hence comparing (2.1.9), (2.2.4) and (2.3.2) we conclude that the MLE's of (θ_1, θ_2) in M_1, $M_1(E)$ and $M_1(L)$ are all different. In general the likelihood equations would be simplest for M_1.

However, the labelled model has some advantages over M_1 or $M_1(E)$. Many classical procedures for detection of outliers can be shown to be likelihood ratio tests in $M_1(L)$. For example, it is easy to show that for $F \sim N(\mu, \sigma^2)$ and $G \sim N(\mu + \delta, \sigma^2)$ $\delta > 0$, in $M_1(L)$, the likelihood ratio test is based on $\tau_{(n)} = (x_{(n)} - \bar{x})/s$ and that for $F \sim \exp(\theta)$, $G \sim \exp(\lambda\theta)$

$\lambda > 1$, the likelihood ratio test is based on Cochran's test $x_{(n)} \Big/ \sum_{i=1}^{n} x_{(i)}$ [cf. Barnett and Lewis (1984)]. From the results of Gather and Kale (1981), we have moreover that these tests are UMP unbiased in $M_1(L)$ for testing $H_0 : \delta = 0$ $(\lambda = 1)$ against $H_1 : \delta > 0$ $(\lambda > 1)$. More generally, Gather and Kale derived UMP-unbiased tests in $M_k(L)$ for the case where F and G are elements of an r-parametric exponential family, where G is stochastically larger than F, and differs only in one component of the parameter.

Another advantage of $M_1(L)$ is that it can be exploited to study the power functions of tests for outliers in the other models M_1 and $M_1(E)$. At least consistency can be checked, since we have seen that as $\delta \to \infty$ (or $\lambda \to \infty$) in the location-shift (or scale inflation models) the limiting behaviour of the order statistics in $M_1(L)$, $M_1(E)$ or M_1 is the same. For instance, Bendre and Kale (1985) have shown that Cochran's test is consistent for $\lambda \to \infty$, i.e., the power of the test tends to one, for any n as $\lambda \to \infty$ under $M_1(L)$ from which we can establish the consistency of Cochran's test in $M_1(E)$ and M_1 as well for $\lambda \to \infty$. For the normal population and shift in location, Bendre and Kale (1985) have shown that the tests based on $\tau_{(n)}$ (Grubb's test), or b_1 (Ferguson's test) and Dixon's test are all consistent in $M_1(L)$ and hence in M_1 and $M_1(E)$ as well. These results also hold for $M_k(L)$ and M_k or $M_k(E)$, when the test statistics are modified in an obvious way.

For $k > 1$, Bendre and Kale (1985) also use $M_k(L)$ to study the masking effect of tests for outliers. This is essentially the study of power functions in $M_{k+r}(L)$ for tests which are originally derived in $M_k(L)$, i.e. when the actual number of outliers is larger than the assumed number k of outliers. Again, asymptotically, the masking effect in M_k or $M_k(E)$ is often the same as that in $M_k(L)$.

We want to point out here that M_k, $M_k(E)$ and $M_k(L)$ all specify the number k of outliers. Another model, which is called identified outlier model $M_k(I)$, assumes moreover that the indexing subset $\nu = (\nu_1, \ldots, \nu_k) \in N_{(k)}$ which labels the outliers, is known. The joint p.d.f. of $(X_1, \ldots X_n)$ is then given by

$$L(\mathbf{x} \mid f, g) = \prod_{i \notin \nu} f(x_i) \cdot \prod_{i \in \nu} g(x_i); \; F \in \mathcal{F}, G \in \mathcal{G} \tag{2.3.3}$$

for some given $\nu \in N_{(k)}$.

This is in fact a two sample set-up, a sample of size $(n - k)$ from F and a sample of size k from G. If we consider the distribution of any symmetric function of the observations in $M_k(I)$, then it is the same as that in M_k or $M_k(E)$. Of course, restriction to symmetric functions of the observations in $M_k(I)$ is equivalent to ignoring that ν is assumed to be known. Mostly, $M_k(I)$ is used as a computational device only, as the assumption that ν is known, is usually unrealistic in the context of a k-outlier problem. However, $M_1(I)$ was assumed by Veale (1975) and also by Gather (1986) for $F \sim$ Exp (θ) and $G \sim$ Exp $(\lambda\theta)$ when estimating θ in outlier situations. See also Barnett and Lewis (1984).

The objections against assuming ν to be known can be raised against the assumption that k, the number of outliers is known which, is common to M_k, $M_k(E)$ and $M_k(L)$. One possibility of not specifying k, is to introduce $k \in \{0, 1, \ldots, n\}$ as an additional parameter in the model. In the context of outlier situations, only small values of k are of interest. Usually k between 2 and 20 per cent of n seems to be the folklore as mentioned in Hampel (1968).

One way of achieving such a percentage p of outliers is to assume a prior binomial distribution $b(n, p)$ for k. This will lead us to the mixture model which was first proposed by Newcomb (1886), Jeffreys (1932), and plays a crucial rule in studies on robustness, one aspect of which is the effect of outliers. There is, of course, a vast amount of literature available on mixtures and inference on parameters in mixtures. We will consider the mixture model in the next section.

6.2.4 The Mixture Model

As pointed out above, the main difference between the k-outlier or model M_k and its variants $M_k(E)$ and $M_k(L)$, and the mixture model, $M(M)$, is that in $M(M)$, the number of outliers k is not specified. It can rather be viewed as an extension of M_k by first treating k as a parameter ranging over $K = \{0, 1, \ldots, n\}$ thus yielding as joint p.d.f. of $X_1, \ldots, X_n$:

$$L(\mathbf{x} \mid f, g, k, \nu) = \{ \prod_{i \in \nu} g(x_i)\}\{ \prod_{i \notin \nu} f(x_i)\}, F \in \mathcal{F}, G \in \mathcal{G}, \nu \in N_{(k)}, k \in K. \tag{1.4.1}$$

Then we give k a prior binomial distribution $b(n, p)$ with $p \in (0, 1)$ and given k, a uniform prior distribution to $\nu \in N_{(k)}$. Integrating out (k, ν), we obtain for the joint p.d.f. of $(X_1, \ldots, X_n)$

$$\begin{aligned} L(\mathbf{x} \mid f, g, p) &= \sum_{k=0}^{n} (1-p)^{n-k} p^k \sum_{\nu \in N_{(k)}} \{ \prod_{i \in \nu} g(x_i)\}\{ \prod_{i \notin \nu} f(x_i)\} \\ &= \sum_{k=0}^{n} \binom{n}{k} p^k (1-p)^{n-k} \left[\sum_{\nu \in N_{(k)}} \frac{1}{\binom{n}{k}} \{ \prod_{i \in \nu} g(x_i)\}\{ \prod_{i \notin \nu} f(x_i)\} \right], \end{aligned} \tag{1.4.2}$$

$$F \in \mathcal{F}, G \in \mathcal{G}, p \in (0,1)$$

Noting that the above assumptions are equivalent to considering i.i.d. r.v.'s $Z_i = (X_1, X_i)$, $i = 1, \ldots, n$ such that $P(Y_i = 1) = p$, $P(Y_i = 0) = 1 - p$, and $P(X_i \leqslant x | Y_i = 0) = F(x)$, $P(X_i \leqslant x | y_1 = 1) = G(x)$, we get that (2.4.2) is the same as the classical mixture model where $Y_1, \ldots, Y_n$ are not observable and where the joint p.d.f. of $(X_1, \ldots, X_n)$ is given by

$$L(\mathbf{x} \mid f, g, p) = \sum_{i=1}^{n} \{(1-p) f(x_i) + p g(x_i)\} \tag{1.4.3}$$

$$F \in \mathcal{F}, G \in \mathcal{G}, p \in (0, 1)$$

It may be noted that since (2.4.3) is the p.d.f. of i.i.d. r.v.'s, each of which is distributed as $H(x) = (1-p)F(x) + pG(x)$, the order statistics of the sample is sufficient for $M(M)$.

The problems of inference about the parameters in $M(M)$ is very complex, and has been considered by many authors. We refer to the monograph by Everitt and Hand (1981) and several references therein. A more up-to-date reference is a book in press by Titterington *et al.* (1986).

We can generalize (2.4.3) to more than one source of contamination, yielding as the joint p.d.f. of $(X_1, \ldots, X_n)$

$$L(\mathbf{x} \mid f, g_1, \ldots, g_r, p_1, \ldots, p_r) = \prod_{i=1}^{n} \left\{ \left(1 - \sum_{j=1}^{r} p_j\right) f(x_i) + \sum_{j=1}^{r} p_j g_j(x_i) \right\},$$

$$F \in \mathcal{F},\ G_j \in \mathcal{G},\ p_i \in (0, 1),\ \sum_{1}^{r} p_i < 1 \tag{2.4.4}$$

However, as in M_k or $M_k(E)$ for $k > 1$, it may not be worthwhile to introduce additional sources for outliers and we will consider (2.4.3) only.

Hampel *et al.* (1986) discuss the use of mixtures in the broad context of robust statistical inference which includes robustness against possible outliers. They point out that the values of p in (2.4.3), that are of interest, are usually those less than 10% or at most 20%, although the usual procedures for estimating the parameters of f and g assume the natural range for p, which is the unit interval.

Barnett and Lewis (1984), as well as Tukey (1960), point out that the mixture model is not suitable for the problem of outliers. Following Tukey (1960), we consider a mixture of two normals $F \sim N(\mu, \sigma^2)$ and $G \sim N(\mu, 9\sigma^2)$. Let I denote the interval $(\mu \pm 2{\cdot}53\sigma)$. Then $P_G[X \notin I] = {\cdot}4010$ and $P_F[X \notin I] = {\cdot}0114$ and if $p = {\cdot}10$ then $P_H[X \notin I] = {\cdot}0491$. Note that the conditional probability that X is from G given that $X \notin I$ is ·80. Thus in a sample of size 30, out of 3 observations expected from G only about 1·20 observations per sample would be outside I while out of 27 observations expected from F only about ·31 observations per sample will lie outside I. Instead of change in scale we now consider shift in location so that $F \sim N(\mu, \sigma^2)$ $G \sim N(\mu + 3\sigma, \sigma^2)$ and take $I = (-\infty, \mu + 2{\cdot}33\sigma)$. Then $P_G[X \notin I] = {\cdot}3726$, $P_F[X \notin I] = {\cdot}0024$ and for $p = {\cdot}10$ $P_H[X \notin I] = {\cdot}03932$ and the conditional probability that X comes from G given that $X \notin I$ would be now ·95. Thus in this case in a sample of size 30 out of 3 observations expected from G only about 1·11 observations per sample would be outside I and out of 27 observations coming from F only about ·70 observations per sample would be outside I. Whether or not such an observation from G would appear to be separated from the main body of the data is another question. Further, in either case there would be more than one observation from G mixed with F and belonging to I. This difficulty is also reflected in the behaviour of the mixture model with respect of the property of outlier proneness (or resistance) of a distribution as defined first by Neyman and Scott (1971) and later modified by

Green (1976). We will consider these models from the viewpoint of outlier proneness (or resistance) in the next section.

6.3. OUTLIER-PRONENESS OF DISTRIBUTIONS

6.3.1. Neyman and Scott's Approach

Neyman and Scott (1971) introduced the concept of outlier prone and outlier resistant families of distributions. Let $X_1, \ldots, X_n$ be i.i.d. r.v.'s with d.f. $F \in \mathcal{F}$ will be called *(d, n)-outlier prone* if for given d and $n \geqslant 3$

$$\sup_{F \in \mathcal{F}} P\{X_{(n)} - X_{(n-1)} > d(X_{(n-1)} - X_{(1)}) \mid F\} = 1. \tag{3.1.1}$$

Denoting the event in the brackets by $E_{d,n}$, we note that $E_{d,n}$ is the critical region of a Dixon-test. Thus, if a family $\mathcal{F}$ is (d, n)-outlier prone, it follows that there is a distribution $F \in \mathcal{F}$, under which, using Dixon test with critical value d, $X_{(n)}$ would be declared as an outlier with probability arbitrarily close to one. The family $\mathcal{F}$ is called *completely outlier prone* if $\sup_{F \in \mathcal{F}} P(E_{d,n}) = 1$ for any $d > 0$ and $n \geqslant 3$. $\mathcal{F}$ is called *(d, n)-outlier resistant* if $\sup_{F \in \mathcal{F}} P(E_{d,n}) < 1$ and *completely outlier resistant* if $\sup_{F \in \mathcal{F}} P(E_{d,n}) < 1$ for all $d > 0$, $n \geqslant 3$. Neyman and Scott (1971) showed that any finite collection of location or/and scale parameter families is a completely outlier resistant family. But the families of Gamma or Log-Normal distributions are completely outlier prone.

We will discuss now the outlier proneness or resistance of the outlier generating models, defined in Section 6.2.

Since the event $E_{d,n}$ is symmetric in $(x_1, \ldots, x_n)$, the outlier proneness property will behave in the same manner for the models M_k and $M_k(E)$. We will consider model M_1 first with joint p.d.f. given in (2.1.1). Following Kale (1977) we define $Z_1 = \min_{i \neq \nu} \{X_i\}$ and $Z_2 = \max_{i \neq \nu} \{X_i\}$ we get

$$P(E_{d,n} \mid M_1) \geqslant P(X_\nu > Z_2(1 + d) - dZ_1 \mid M_1)$$
$$= \iint [1 - G(z_2(1 + d) - dz_1)]h(z_1, z_2)\, dz_1, dz_2$$

where $h(z_1, z_2) = (n - 1)(n - 2)[F(z_2) - F(z_1)]^{n-2} f(z_1)f(z_2)$, $z < z_2$. For the shift in location case (2.1.2) for $\delta \to \infty$, we can show that $P(E_{d,n} \mid M_1) \to 1$ for any d. Hence, in this case, M_1 is completely outlier prone.

Similarly, if we consider (2.1.3), the change in scale case, along with the assumption that $F(0) = 0$, then we can show that M_1 is completely outlier prone.

As noted by Kale (1977), the Neyman-Scott definition of outlier-proneness has to be modified if one is looking for k possible outliers.

Instead of $E_{d,n}$, we then have to consider the event

$$\{X_{(n)} - X_{(n-k)} > d(X_{(n-k)} - X_{(1)})\}.$$

For instance, if we use the definition of outlier proneness given in (3.1.1) to study the behaviour of M_2, we can show that M_2 is not generally outlier prone. This phenomenon is connected with the masking effect of the Dixon test where a large value of $X_{(n-1)}$ masks $X_{(n)}$ which results into not rejecting $X_{(n)}$ as an outlier. For details we refer to Bendre and Kale (1985).

Kale (1977) has shown that if $\mathcal{F}$ and $\mathcal{G}$ are each completely outlier-resistant, then so is the family of $(1-p)F + pG$, $F \in \mathcal{F}$, $G \in \mathcal{G}$, $0 \leqslant p \leqslant 1$. Further, if at least one of $\mathcal{F}$ and $\mathcal{G}$ is completely outlier prone, then the above family of mixtures is completely outlier prone. But, the family of distributions of mixtures with p bounded away from zero and one, is outlier resistant if at least one of $\mathcal{F}$ and $\mathcal{G}$ is outlier resistant. Thus the mixtures of Normal distributions are outlier resistant. Another consequence of Neyman and Scott's definition is the result that every location or/and scale family is completely outlier resistant, even the family of all Cauchy-distributions. Both consequences have resulted into casting doubts on whether Neyman and Scott's definition is suitable to describe the outlier proneness of distributions.

Another criticism is the fact that no single distribution and even no finite set of distributions can be completely outlier prone [Green (1974), Gather (1979)]; this follows from Green's (1974) result that (d, n)-outlier proneness for some $d > 0$, $n \geqslant 3$ and complete outlier proneness are equivalent.

Green (1976) substantially changed the viewpoint by defining outlier proneness not for a family of distributions but for a given single c.d.f. F and by considering the asymptotic behaviour of $X_{(n)}$, $X_{(n-1)}$ from F as $n \to \infty$.

6.3.2. Green's Approach

Though similar at first glance, outlier resistance in the sense of Green (1976) is quite different from that in the sense of Neyman and Scott, discussed in Section 6.3.1. Green's definition allows us to speak of outlier resistance or outlier proneness of a single d.f. F. Based on asymptotic properties of $X_{(n)}$ and $X_{(n-1)}$, it is closely related to the Weak Law of Large Numbers for the extremes. We will assume that $\{X_i\}_{i=1}^{\infty}$ is a sequence of i.i.d. r.v.'s with d.f. F such that $F(x) < 1$ for any $x \in R$.

Green called the d.f. F *absolutely outlier resistant* (AOR) if for any $\epsilon > 0$

$$P(X_{(n)} - X_{(n-1)} > \epsilon) \to 0, \text{ as } n \to \infty. \tag{3.2.1}$$

F is defined to be *relatively outlier resistant* (ROR) if for any $\gamma > 1$

$$P(X_{(n)} > \gamma X_{(n-1)}) \to 0, \text{ as } n \to \infty. \tag{3.2.2}$$

These definitions reflect the idea that in large samples from an absolutely (relatively) outlier resistant distribution, the gap (ratio) between the two largest observations should be small with probability close to one. Green did not define absolute (relative) outlier proneness by a simple negation of absolute (relative) outlier resistance, but by requiring that F be called absolutely (relatively) outlier prone if, even for infinitely large samples, the gap (ratio) between the two largest observations is bounded away from zero (one) with positive probability. Thus, F is *absolutely outlier prone* (AOP), if there exist ϵ_0, $\delta_0 > 0$ and n_0 such that

$$P(X_{(n)} - X_{(n-1)} > \epsilon_0) \geqslant \delta_0 \quad \text{for all } n \geqslant n_0 \tag{3.2.3}$$

and F is called *relatively outlier prone* (ROP) if there exist $\gamma_0 > 1$, $\delta_0 > 0$, and n_0 such that

$$P(X_{(n)} - \gamma_0 X_{(n-1)}) \geqslant \delta_0 \quad \text{for all } n \geqslant n_0. \tag{3.2.4}$$

Using classical results of Gnedenko (1943) and Geffroy (1958) about the equivalence of (3.2.1) and (3.2.2) with the additive and multiplicative WLLN for extremes, Green established the following necessary and sufficient conditions for F to be absolutely (relatively) outlier resistant or -prone.

Theorem 3.2.1. (a) F is absolutely outlier resistant (AOR) if for any $\epsilon > 0$, $\lim\limits_{x \to \infty} \dfrac{1 - F(x + \epsilon)}{1 - F(x)} = 0.$

(b) F is relatively outlier resistant (ROR) if for any $\gamma > 1$,

$$\lim_{x \to \infty} \frac{1 - F(\gamma x)}{1 - F(x)} = 0.$$

Thus the outlier resistance is characterized by the limiting behaviour of the ratio of the tail probabilities of F. Green also proved a similar characterization for outlier proneness.

Theorem 3.2.2. (a) F is absolutely outlier prone (AOP) if there exist ϵ_0, $\delta_0 > 0$ and that $\dfrac{1 - F(x + \epsilon_0)}{1 - F(x)} \geqslant \delta_0$ for all $x \in R$.

(b) F is relatively outlier prone (ROP) if there exist $\gamma_0 > 1$, $\delta_0 > 0$ such that $\dfrac{1 - F(\gamma_0 x)}{1 - F(x)} \geqslant \delta_0$ for all $x \in R$.

One can obtain sufficient conditions for AOR (ROR) and AOP (ROP) based on ratios of p.d.f.'s $\dfrac{f(x + \epsilon)}{f(x)}$ and $\dfrac{f(\gamma x)}{f(x)}$, which in some special cases are also necessary conditions.

From the above theorems, we have that if F is AOR then it is ROR. Similarly if F is ROP then it is AOP. Of course F cannot be simultaneously AOR and AOP and similarly it cannot be simultaneously ROP and ROR. On the other hand as demonstrated by Green (1976) all these four categories are not mutually exclusive. Thus, these four definitions allow a fine differentiation of distributions with respect to their outlier behaviour. For example, using the characterization theorems, one can easily check that any normal distribution is AOR, and any exponential distribution is AOP but ROR. Any Cauchy distribution is ROP and thus also AOP. Note that the family of Cauchy distributions is not outlier prone in the sense of Neyman and Scott, which was an important point in the criticism against that approach.

The definitions of absolute (relative) outlier resistance and proneness need not be modified when we are looking for possible k outliers, since the probabilities $P(X_{(n)} - X_{(n-k)} > \epsilon)$ and $P(X_{(n)} > \gamma X_{(n-k)})$ show up the same asymptotic behaviour for $n \to \infty$ as those in (3.2.1) and (3.2.2) if k is a fixed given integer, not depending on n [cf. Geffroy (1958)].

Gather (1979) showed that if $F(x) = 1 - e^{-xa(x)}$ for $x \geqslant x_0$ then F is AOR if $a(x)$ is increasing and $a(x) \to \infty$, $x \to \infty$. If $a(x)$ is bounded then F cannot be AOR. Similarly, she showed that distributions with densities of Weibull-type $f(x) = cx^{p-1} \exp(-bx^{\alpha})$, $x > 0$, $p > 0$, $b > 0$, $\alpha > 0$ are AOR if $\alpha > 1$ and ROR for all $\alpha > 0$.

As in subsection 6.3.1, we now consider how the various outlier generating models (presented in Section 6.2) compare with respect to outlier proneness and resistance with the homogeneous case H_0. The question we try to answer is, are these models more AOP (ROP) then the model H_0. More precisely we are interested in the behaviour of

$$P(X_{(n)} - X_{(n-1)} > \epsilon \mid M),\ \epsilon > 0 \tag{3.2.5}$$

and,

$$P(X_{(n)} > \gamma X_{(n-1)} \mid M),\ \gamma > 1 \tag{3.2.6}$$

where M is either M_k, $M_k(E)$, $M_k(L)$ or $M(M)$. We will concentrate on the case $k = 1$ and we also note that M_k and $M_k(E)$ need not to be discussed separately as the behaviour of (3.2.5) and (3.2.6) is the same under both models.

Gather (1984) showed that if in (2.1.1), (2.2.1), (2.3.1), (2.4.3), the d.f. G, that is the d.f. from which the outlier is supposed to result, is AOP (ROP), then the models M_1, $M_1(E)$, $M_1(L)$, $M(M)$ are also AOP (ROP).

More precisely, there exist $\epsilon_0 > 0 (\gamma_0 > 1) \delta_0 > 0$, and n_0 such that

$$P(X_{(n)} - X_{(n-1)} > \epsilon_0 \mid M) \geqslant \delta_0 \quad \text{for all } n \geqslant n_0$$

$$(P(X_{(n)} > \gamma_0 X_{(n-1)} \mid M) \geqslant \delta_0 \quad \text{for all } n \geqslant n_0)$$

if M is either M_1, $M_1(E)$, $M_1(L)$ or $M(M)$.

She also showed that all these models are AOR (ROR) in the above sense whenever F and G are both AOR (ROR). Thus, in particular, for the shift of location case with $G(x) = F(x - \delta)$, and the change of scale case where $G(x) = F(x/\lambda)$, all the models M_1, $M_1(E)$, $M_1(L)$ and $M(M)$ are AOR (ROR) if F itself is AOR (ROR) and a similar remark holds true whenever both F and G are AOP (ROP).

For example, if F is exponential with mean σ, thus known to be ROR, then the models M_1, $M_1(E)$, $M(M)$, $M_1(L)$, obtained by taking G exponential with mean $\lambda\sigma$, $\lambda > 1$, are all ROR. That is, the outlier generating models are ROR whenever corresponding homogeneous model H_0 is ROR. Hence, the interesting question is what happens if the order of convergence in (3.2.2) is changed.

Gather (1984) proved that for the mixture model $M(M)$ this order does not change whereas for the labelled model $M_1(L)$, the probability $P(X_{(n)}, > \gamma X_{(n-1)})$ in (3.2.2) tends to zero more slowly than for H_0 for any $\gamma > 1$. On the other hand, in M_1 or $M_1(E)$ the rate does not change for $1 < \gamma \leqslant \frac{\lambda}{\lambda - 1}$ but (3.2.2) holds with slower rate than for H_0 if $\gamma > \frac{\lambda}{\lambda - 1}$. It is interesting to note that if the mixing proportion is chosen to be $p = \frac{1}{n}$, then $M(M)$ shows the same behaviour as M_1 or $M_1(E)$.

Similar results were proved by Gather (1984) for some other distributions, as Normals, Log-Normals and Weibull-distributions. Mostly, the exact order of convergence of (3.2.2) was given as well.

From the above results we may deduce a hierarchy between the outlier generating models with respect to their degree of outlier resistance. Thus, in the exponential case if $\lambda \leqslant \gamma/\gamma - 1$ (i.e. a small scale inflation for the outlier observation) the models M_1, $M_1(E)$ and $M(M)$ have the same degree of outlier resistance as the homogeneous model H_0, where no outliers are present. However, if the scale inflation is large, then M_1 and $M_1(E)$ are less outlier resistant than the mixture model, and H_0, the homogeneous model. Considering the orders of convergence in (3.2.2), $M_1(L)$ has the slowest rate, and thus is the least outlier resistant among the four models considered here. Hence, $M_1(L)$ is maximally separated from the homogeneous model H_0.

However, a criticism against Green's approach is that it depends on the asymptotic behaviour of $X_{(n)}$, $X_{(n-1)}$ as $n \to \infty$, whereas in the context of outliers, the main interest is in samples of moderate size with a small number of outliers. Another criticism might be that the four categories given by AOR, ROR, ROP, AOP are not all mutually exclusive. This results in a possibly too fine distinction into different types of outlier behaviour.

However, Cauchy-distributions are relatively and absolutely outlier prone in Green's sense, a property which Neyman-Scott's definition did not imply and which is regarded as a serious deficiency of that approach.

It is interesting to note that according to both approaches, the mixture model, with mixtures of a distribution $F(x)$ with $F(x - \delta)$ or $F(\lambda x)$ is neither outlier prone nor less outlier resistant than H_0, if F itself is not outlier prone.

However, Green's approach yields that mixtures of F and G, with G AOP (ROP), are also AOP (ROP), while the analogous property is not true for Neyman-Scott's approach, where classes of mixtures with at least one outlier resistant part remain outlier resistant.

Some modifications of Green's approach have been considered. Realizing that no discrete distribution can be AOR, Gather and Mathar (1983) introduced the concept of d-absolute-, and q-relative outlier resistance and proneness, based on the asymptotic behaviour of, for example,

$$P(X_{(n)} - X_{(n-1)} > \epsilon + d),\ d \geqslant 0,\ \epsilon > 0$$

and

$$P(X_{(n)} > q\gamma X_{(n-1)}),\ q \geqslant 1,\ \gamma > 1.$$

A definition based on the limit behaviour of Dixon's statistic is investigated in Gather (1981) and one based on the upper record values X_{U_n} and $X_{U_{n-1}}$ is studied in Gather and Pfeifer (1986).

One main concern about the concepts of outlier resistance and -proneness as given above is whether they have reasonable consequences in statistical framework. If $\mathcal{F} = \{F(x - \theta);\ \theta \in R\}$ is a location family satisfying some smoothness conditions, we have from Gather (1984, 1986b), that if F is ROP, then the trimmed means $Z_{n;\,r} := \frac{1}{n-r} \sum_{i=r+1}^{n-4} X_{(i)}$, $0 \leqslant r \leqslant \frac{n-1}{2}$, show up a worst-case behaviour in the sense of Jureckova (1981) who used

$$B(a, T_n) := \frac{-lnP_\theta(|T_n - \theta| > a)}{-lnP_\theta(|X_1 - \theta| > a)}$$

[cf. Bahadur (1967)] for $a \to \infty$ as a measure of performance of an estimator $T_n = T_n(X_1, \ldots, X_n)$ of θ. This result restates in particular that the arithmetic mean $Z_{n:0}$ is a bad estimator of location for heavy tailed distributions. It is an open question if for ROR distributions, $B(a, Z_{n:r})$, as $a \to \infty$, attains its upper bound; this is however shown by Jureckova (1981) for some special light-tailed distributions including the normal.

Mathar (1985) proved that an estimator $V_n := \sum_{i=0}^{r} \alpha_{i,n} X_{(n-i)} - a_n$, where $\alpha_{i,n} \geqslant 0$, $\sum_{i=1}^{r} \alpha_{i,n} = 1$, $a_n \in R$, which is based on the r upper order statistics is consistent for θ if and only if F is AOR.

Mathar (1985) also proved the following implication of the AOR property of F on the separability of the simple hypotheses $H_0 : \theta = \theta_0$ *vs* $H_1 : \theta = \theta_1$, $\theta_1 > \theta_0$. If ρ^*_α denotes the MP test for H_0 *vs* H_1 at level α we have that

$$\lim_{\alpha \to 0^+} \sup \frac{E_{\theta_0}(\rho^*_\alpha)}{E_{\theta_1}(\rho^*_\alpha)} = 0$$

if and only if F is AOR, which means that the type II error-probability converges to zero more slowly than the type I error-probability thus yielding even for small level α a separation of H_0 and H_1 with fairly high power if F is AOR.

6.4 BAYESIAN APPROACH

A Bayesian approach to the outlier problem has been implicit in the exchangeable model as well as in the mixture model and a trace of such an approach is present in Pierce's (1852) argument also. One of the earliest discussions of a Bayesian approach to outlier rejection is that of de Finetti (1961) although Box and Tiao (1968) appear to be the first ones to present an extensive analysis of Bayesian inference in the presence of outliers. In general they assign a prior distribution $p_{(k)}$ to the 2^n sets $\nu \in N_{(k)}$, $k = 0, 1, 2, \ldots, n$ as well as prior distributions for the parameters occurring in F and G in the model given by (2.1.10). Thus for $F \sim N(\mu, \sigma^2)$ and $G \sim N(\mu, \lambda^2\sigma^2)$ with $\lambda^2 \geqslant 1$ assumed fixed, they obtain the posterior distribution of μ assuming local uniform priors for μ and $\log \sigma$. More specifically they recommend $p_{(k)} = p^k(1 - p)^{n-k}$ for $k = 0, 1, 2, \ldots, n$ thus leading to $M_k(E)$ first and to $M(M)$ later. Dempster and Rosner (1971) using $M_k(E)$ considering normal shift in location problems, and using local uniform priors over $(\mu, 1/\sigma^2)$ obtain the posterior probability distribution $P(\nu)$, $\nu \in N_{(k)}$, $k = 0_1, \ldots, n$, and show that the observations with largest absolute values have maximum posterior probability of coming from the distribution G. As mentioned in subsection 6.2.2 Kale (1976) proved a similar result for $M_k(E)$ where F and G belong to a one parameter exponential class of densities and uniform priors were given to $\nu \in N_{(k)}$, for fixed k, leading to the result that the k largest (smallest) observations have maximum posterior probability of resulting from G. Guttman (1973) considering $M_1(E)$ for shift in location in the normal distribution and assuming priors for (μ, δ, σ^2) proportional to $1/\sigma^2$ obtained the posterior distribution of δ. Guttman and Khatri (1975) generalized these results for $M_1(E)$ and $M_2(E)$ with normal distributions and shift in location or/and scale.

Kale (1969) using a similar technique obtained (semi) Bayesian estimator of θ, in case of $M_1(E)$ where F is exp (θ), G is exp $(\lambda\theta)$, $\lambda \geqslant 1$ and λ has a beta distribution of second kind or equivalently $1/\lambda$ has a Beta distribution of first kind. Sinha (1972) gave a Baye's estimator of the reliability function in the above set up.

As observed in Section 6.2, the outlier generating models are extensions of the model H_0 of no outliers. Box and Tiao [(1964), (1973)] recommend extending model H_0 with $F \sim N(\mu, \sigma^2)$ to

$$f(x, \mu, \sigma^2, \beta) = \frac{c}{\sigma} \exp\left\{\frac{1}{2}\left|\frac{x-\mu}{\sigma}\right|^{2/(1+\beta)}\right\} -1 < \beta \leqslant 1.$$

For $\beta = 0$ we get normal distributions and with $\beta = 1$ we get double exponentials, while for $\beta = -1$, the distribution tends to be uniform over $(\mu \pm \sqrt{3}\bar{\sigma})$ Smith (1983) has recommended using this family for possible outliers, as the family includes distributions with thicker tails. Smith (1983) also advocates the use of a prior distribution $p_{(k)}$ over $\nu \in N_{(k)}$ and $k = 0, 1, 2, \ldots, n$ such that $\sum_{k \geqslant 3} p_{(k)}$ is negligible. This leads essentially to M_0, M_1 or M_2 models considered in Section 6.2, and Smith (1983) also shows that it is natural to suspect the extreme observations in this case.

Following de Finetti's argument that any approach to rejection of outliers needs to be based on the effect of possible outlying observations on the posterior distribution, O'Hagan (1979) and Goldstein [(1982), (1983)] define outlier proneness of distributions on the basis of the influence one large observation may have on the posterior. Consider for example a location parameter family $\mathcal{F} = \{F(x-\theta),\ \theta \in R\}$. Then $\mathcal{F}$ is called *B-right outlier prone of order n* if for all $c, x_1, \ldots, x_n$, and all prior distributions on θ, we have

$$\lim_{X_{n+1}\to\infty} P(\theta \leqslant c \mid X_1 = x_1, \ldots, X_n = x_n, X_{n+1} = x_{n+1}) \\ = P(\theta \leqslant c \mid X_1 = x_1, \ldots, X_n = x_n) \tag{4.1.1}$$

where $X_1, \ldots, X_{n+1}$ is a sample from F. *B-left outlier proneness* is defined analogously, and is called *B-outlier prone of order n* if it is both B-right and B-left outlier prone. $\mathcal{F}$ is called *B-outlier resistant* if for all $c, x_1, \ldots, x_n$ and all prior distributions on θ

$$P(\theta \leqslant c \mid X_1 = x_1, \ldots, X_n = x_n, X_n = x_{n+1}) \tag{4.1.2}$$

is strictly decreasing in x_{n+1}.

For bounded densities f, outlier proneness of order n implies outliers proneness of order $n+1$, and thus outlier proneness of order one is the strongest property. It seems that characterizations of (4.1.1) and (4.1.2) similar to those of Green in Theorems 3.2.1 and 3.2.2 above, are not possible. But O'Hagan (1979) has given the following sufficient conditions: (i) f is a continuous, symmetric density which is decreasing for $x \geqslant x_0$, (ii) $\partial \log f(x)/x$ exists and is decreasing for $x \geqslant x_0$, and (iii) for any $\epsilon > 0$ there exist $\delta > 0$, $x_1 \in R$ such that

$$\left|\frac{f(x')}{f(x)} - 1\right| < \delta \text{ for all } x, x' \geqslant x_1, \text{ and } |x - x'| < \epsilon \tag{4.1.3}$$

Under the above conditions (i), (ii), and (iii) $\mathcal{F}$ is B-outlier prone of order 1. Similarly, if $\log f(x)/x$ exists for all $x \in R$ and is decreasing for $x \geqslant x_0$, say, then the boundedness of f' implies that $\mathcal{F}$ is B-outlier resistant. Apparently there are no simple relationships between Neyman Scott's, Green's and O'Hagan's approach to outlier proneness. Although, as shown by Gather (1984), a distribution function F which satisfies the sufficient conditions (4.1.3) for B-outlier proneness, is also absolutely outlier prone in the sense of Green (3.2.1). But, the sufficient conditions for B-outlier resistance are neither sufficient nor necessary for absolute outlier resistance in Green's sense. Counter example is given by Gather (1984), where

$$f(x) = c \exp[-|x| + \log |x|],\ |x| > 2$$
$$= c \exp\left[\frac{1}{128}x^4 - \frac{1}{16}x^2 - \frac{13}{8} - \log 2\right],\ |x| \leqslant 2$$

is B-outlier resistant because of (4.1.3), but is not AOR, and on the other hand, a distribution with density $f(x) = ce^{-x^2}|\sin |x||$ is AOR but does not satisfy the sufficient condition (4.1.3) for B-outlier resistance.

The main difference between Green's and O'Hagan's approach is that for Green's property (3.2.1), the limit of $f(x+\epsilon)/f(x)$ for $x \to \infty$ is considered (if densities exist and are smooth enough), whereas O'Hagan has to check the monotonicity of $f(x+\epsilon)/f(x)$. However, both definitions will yield that distributions with heavy tails, such as Cauchy, are outlier prone and that light-tailed distributions such as normal are outlier resistant.

The outlier proneness in Bayesian sense has not been investigated fully. For example, the problem of whether the 1-outlier model M_1 is B-outlier prone or not remains an open question, and there is a lot of scope for additional work.

6.5. GENERAL COMMENTS

Among the models discussed in Section 6.2, the authors feel that there are many good arguments to support the k outlier model M_k or possibly a union of a few M_k for sufficiently small values of k, say less than or equal to about 20% of n, the sample size. Although the order statistics are not sufficient for M_k, as seen earlier, the restriction to procedures based on order statistics can be justified on the grounds of either invariance or conditional likelihood. Further under mild conditions on G and F, one can also justify suspecting extreme observations as outliers and it is not difficult to develop the theory of maximum likelihood estimation of parameters of F and G as well as k, regarded as an unknown parameter. Moreover, if the object is to detect and/or identify the outliers then some of the well-known existing procedures turn out to be optimal ones in M_k.

In the exchangeable model $M_k(E)$, the order statistics are sufficient and although the detection and labelling of outliers is similar to that in M_k, the MLES of parameters of F, G are easier to obtain in M_k than in $M_k(E)$. In the labelled model $M_k(L)$ further sufficient reduction of data is possible as it is assumed that the extreme observations are in fact outliers coming from G. $M_k(L)$ thus can be viewed as a conditional model (or truncated model) arising out of M_k, where the outliers are identified with extreme order statistics. The classical test procedures for detecting outliers turn out to be likelihood ratio tests and are even UMP unbiased in some specific cases. Also, $M_k(L)$ turns out to be more outlier prone than all other models with respect to Green's concept of outlier proneness. Finally, $M_k(L)$ provides good computational device for studying the masking effect and the consistency of several outlier tests under model M_k. However, in M_k the probability that extreme observations are in fact outliers could be much smaller than one, even when G is far removed from F. A more detailed study of the effect of restricting M_k to $M_k(L)$ either viewed as truncated or conditioning would be extremely useful.

The mixture model $M(M)$ perhaps is not very suitable as even very small values of p, the mixing proportion still contribute to the terms of the type $p^k(1-p)^{n-k}$ even for values of k as large as n, at least in theoretical derivation of the properties of this model. Thus we still do not know, even in the case where $\partial G/\partial F$ is monotone whether extreme observations have higher probability (conditional or otherwise) of observations coming from G. In fact although mixtures can be generated by a mechanism which allows one to label X_i as coming from G when in fact X_i comes from $H = (1-p)F + pG$. Another problem with the mixture model is that it behaves differently from the other three models in relation to the outlier proneness of distributions in the sense of Neyman and Scott (1971) or Green (1976). Thus although M_k, $M_k(E)$, $M_k(L)$ are in general more outlier prone than the model for the homogeneous case H_0, the same remark does not hold for $M(M)$. Of course one could regard this as a defect of the Neyman-Scott and/or Green's approach rather than a defect of the mixture model. However much more work is needed before this issue can be resolved satisfactorily. Needless to add that modelling is in general not an easy task and for outliers which is essentially a post data concept the job becomes much harder. It is hoped that this review paper will stimulate further interest in these aspects of the old and interesting problem of outliers.

REFERENCES

1. Airy, G. B. (1856): Letter from Professor Airy, Astronomer Royal to the editor, *The Astronomical Journal*, **4**, 137–138.
2. Anscombe, F. J. (1960): Rejection of Outliers, *Technometrics*, **2**, 123–147.
3. Anscombe, F. J. and Barron, B. A. (1966): Treatment of Outliers in samples of size three, *Jour. Res. Nat. Bur. Standards*, **B, 70**, 141–147.
4. Bahadur, R. R. (1967): Rates of convergence of estimates, *Ann. Math. Statist*, **38**, 303–324.
5. Barnett, V. and Lewis, T. (1978): *Outliers in Statistical Data*, John Wiley and Sons, New York.
6. ——— (1984): *Outliers in Statistical Data*, John Wiley and Sons, New York.
7. Bendre, S. M. (1985): Masking and Swamping Effects on Test for Outliers, Ph.D. thesis, University of Poona, Pune, India.
8. Bendre, S. M. and Kale, B. K. (1985): Masking effect on tests for discordant observations in normal samples, Tech. Report, 73, Dept. of Statistics, University of Poona, Pune, India.

9. ——— (1985): Masking effect on tests for outliers in exponential models, *Jour. Amer. Stat. Assn.*, **80**, 1020–1025.

10. Bessel, F. W. and Baeuer, J. J. (1838): *Gradmessung in Ostpreussen*, reprinted in *Abhandlungen von F. W. Bessel*, R. Engleman (Editor), Leipzig, Germany, 1876.

11. Box, G. E. P. and Tiao, G. C. (1964): A Bayesian approach to the importance of assumptions applied to the comparisons of variances, *Biometrika*, **51**, 153–167.

12. ——— (1968): A Bayesian approach to some outlier problems, *Biometrika*, **55**, 119–129.

13. ——— (1973): *Baysian Inference in Statistical Analysis*, Addison-Wesley, London.

14. Chauvenet, W. (1863): Method of Least Squares, Appendix to *Manual of Spherical and Practical Astronomy*, **Vol. 2**, Philadelphia: Lippincott.

15. Coolidge, J. L. (1925): *An Introduction to Mathematical Probability*, Oxford University Press, England.

16. Cox, D. R. (1975): Partial Likelihood, *Biometrika*, **62**, 269–276.

17. David, H. A. and Shu, V. S. (1978): *Robustness of location estimators in the presence of an outlier, Contributions to Survey Sampling and Applied Statistics in Honour of H. O. Hartley*, H. A. David (Editor), Academic Press, New York.

18. David, H. A. (1979): *Robust estimation in the presence of outliers, Robustness in Statistics*, R. L. Launer and G. N. Wilkinson (Editors), Academic Press, New York.

19. ——— (1981); *Order Statistics*, Second Edition, John Wiley and Sons, New York.

20. ——— (1985): *Order statistics under non-standard conditions, Biostatistics: Statistics in Biomedical, Public Health and Environmental Science*, P. K. Sen (Editor), Elsevier Science Publishers, Amsterdam, Holland.

21. Dempster, A. P. and Rosner, B. (1971): *Detection of Outliers, Statistical Decision Theory and Related Topics*, S. S. Gupta and J. Yackel (Editors), Academic Press, New York.

22. Dixon, W. J. (1950): Analysis of extreme values, *Ann. Math, Stat.*, **21**, 488–506.

23. ——— (1951): Ratios involving extreme values, *Ann. Math. Stat.*, **22**, 68–78.

24. ——— (1953); Processing data for outliers, *Biometrics*, **9**, 74–89.

25. Edgeworth, F. Y. (1887): On discordant observations, *Phil. Mag.*, **23**, 364–375.

26. Everitt, B. and Hand, D. J. (1981): *Finite Mixtures of Distributions*, Chapman and Hall, London, England.

27. Ferguson, T. S. (1961): On the rejection of outliers, Proc. of 4th Berkley Symp., Prob. Stat., **1**, 253–287.

28. de Finetti, B. (1961): The Bayesian approach to the rejection of outliers, Proc. of 4th Berkley Symp., *Prob. Stat.*, **1**, 199–210.

29. Gather, U. (1979): On tests for outliers and outlier proneness of distributions, Ph.D. Thesis (in German), Technical University of Aachen.

30. ——— (1981): A characterization of the outlier behaviour of distributions based on Dixon's statistics (in German), *Meth. Oper. Res.*, **44**, 81–90.

31. Gather, U. and Kale, B. K. (1981): UMP—tests for *r*-upper outliers in samples from exponential families, Proc. Indian Stat. Inst. Golden Jubilee, Int. Conf. on Statistics: Applications and New Directions, J. K. Ghosh *et al.* (Editor), Indian Statistical Institute, Calcutta, India.

32. Gather, U. and Helmers, M. (1983): A locally most powerful test for outliers in samples from the exponential distribution, *Math. Oper. Res.*, **47**, 39–47.

33. Gather, U. and Mathar, R. (1983): Analyzing the outlier—behaviour of non-continuous distribution functions, *Jour. Ind. Stat. Assoc.*, **21**, 9–18.

34. Gather, U. (1984): Tests and estimators in outlier models, Habilitation Thesis (in German), Technical University of Aachen.

35. ——— (1985): The influence of outlier-proneness on the tail-behaviour of some location estimators, *Statistics and Decisions*, **Suppl. Iss. 2**, 165–171.

36. ——— (1986a): Robust estimation of the mean of the exponential distribution in outlier situations. To appear in Comm. Statist., Tech. Report, No. **115**, Dept. of Statistics, University of Iowa.

37. ——— (1986b): On counterparts to the Weak Law of Large Numbers for extremes. Tech. Report, **No. 116**, Dept. of Statistics, University of Iowa.

38. Gather, U. and Pfeifer, D. (1986): Maxima, records and stability. Submitted to Appl. Prob., Tech. Report, No. **114**, Dept. of Statistics, University of Iowa.

39. Geffroy, J. (1958): *Contributions à la théorie des valeurs extrèmes, Publ. Inst. Stat.*, University of Paris, **7**, **8**, 37–184.

40. Gnedenko, B. V. (1943): *Sur la distribution limite du terme maximum d' une série aléatoire*, *Ann. Math.*, **44**, 423–453.

41. Godambe, V. P. (1976): Conditional likelihood and unconditional optimum estimating equations, *Biometrika*, **63**, 277–284.

42. Goldstein, M. (1982): Contamination distributions, *Ann. Stat.*, **10**, 174–178.

43. ——— (1983): Outlier-resistant distributions: where does the probability go?, *Jour. Roy. Stat. Soc.*, **B**, **45**, 355–357.

44. Green, R. F. (1974): A note on outlier-prone families of distributions, *Ann. Stat.*, **2**, 1293–1295.

45. ——— (1976): Outlier-prone and outlier-resistant distributions, *Jour. Amer. Stat. Assoc.*, **71**, 502–505.

46. Grubbs, F. E. (1950): Sample criteria for testing outlying observations, *Ann. Math. Stat*, **21**, 27–58.

47. Guttman, I. and Smith, D. E. (1969): Investigation of rules for dealing with outliers in small samples from normal distribution, I: Estimation of the mean, *Technometrics*, **11**, 527–550.

48. Guttman, I. (1973): Care and handling of univariate or multivariate outliers in detecting spuriosity—a Bayesian approach, *Technometrics*, **15**, 723–738.

49. Guttman, I. and Khatri, C. G. (1975): A Bayesian approach to some problems involving the detection of spuriosity, *Applied Statistics*, R. P. Gupta (Editor), Amsterdam, North Holland.

50. Hall, I. J. and Kudo, A. (1968): On slippage tests (I), a generalization of the Neyman-Pearson Lemma, *Ann. Math. Stat.*, **39**, 1693–1699.

51. Hampel, F. R. (1968): Contributions to the theory of robust estimation, Ph.D. Thesis, University of California, Berkeley, California.

52. Hawkins, D. M. (1980): *Identification of Outliers*, Chapman and Hall, London, England.

53. Jeffreys, H. (1932): An alternative to the rejection of observations, Proc. Roy. Soc., London, **A**, **137**, 78–87.

54. ——— (1938): The law of error and combination of observations, Phil. Trans. Roy. Soc., London, **A**, **237**, 231–271.

55. Jurecková, J. (1981): Tail-behavior of location estimators, *Ann. Stat.*, **9**, 578–585.

56. Kale, B. K. (1969): Estimation of expected life in the presence of an outlier observation, a Bayesian approach. Contributed paper, 37th Session of Int. Stat. Inst. meeting, London, England, 150–152.

57. Kale, B. K. and Sinha, S. K. (1971): Estimation of expected life in the presence of an outlier observation, *Technometrics*, **13**, 755–759.

58. Kale, B. K, (1975): Trimmed means and the method of MLE when spurious observations are present, *Applied Statistics*, R. P. Gupta (Editor), Amsterdam, Holland.

59. ——— (1976): Detection of Outliers, *Sankhya*, **B**, **38**, 356–363.

60. ——— (1977): A note on the outlier-resistant families and mixture of distributions, *Jour. Ind. Stat. Assoc.*, **15**, 119–130.

61. Kudo, A. (1956): On testing outlying observations, *Sankhya*, **17**, 67–76.

62. Legendre, A. M. (1805): On the method of least squares, Translated from French in *A Source Book in Mathematics*, D. E. Smith (Editor). Dover Publications.

63. Mathar, R. (1985): Laws of large numbers and the outlier behavior of dependent random variables (in German). Habilitation Thesis, Technical University of Aachen.

64. Mount, K. S. and Kale, B. K. (1973): On selecting a spurious observation, *Can. Math. Bulln.*, **16**, 75–78.

65. Murphy, R. B. (1951): On tests for outlying observations. Ph.D. Thesis, Princeton University, Princeton, New Jersey.

66. Newcomb, S. (1886): A generalized theory of the combination of observations so as to obtain the best result, *Amer. Jour. Math.*, **8**, 343–366.

67. Neyman, J. and Scott, E. L. (1971): Outlier proneness of phenomena and of related distributions, *Optimizing Methods in Statistics*, J. Rustagi (Editor). New York: Academic Press.

68. O'Hagan, A. (1979): On outlier rejection phenomena in Bayes inference, *Jour. Roy. Stat. Soc.*, **B**, **41**, 358–367.

69. Paulson, E. (1952): An optimum solution to the *k*-sample slippage problem for the normal distribution, *Ann. Math. Stat.*, **23**, 610–616.

70. Pearson, E. S. and Chandrasekar, C. (1936): The efficiency of statistical tools and a criterion for the rejection of outlying observations, *Biometrika*, **28**, 308–320.

71. Peirce, B. (1852): Criterion for the rejection of doubtful observations, *Astronomical Journal*, **2**, 161–163.

72. Rauhut, B. O. (1982): Estimation of the mean of the exponential distribution with an outlying observation, *Comm. Stat.*, **A**, **11**, 1439–1452.

73. Sinha, S. K. (1973): Lifetesting and reliability estimation for non-homogeneous data, a Bayesian approach; *Comm. Stat.*, **2**, 235–243.

74. Small, C. G. and McLeish, D. L. (1985): Maximum likelihood estimation for scale-slippage outlier models, Submitted to *Technometrics*.

75. Smith, A. F. M. (1983): *Bayesian approaches to outliers and robustness, Specifying Statistical Models: From Parametric to Non-parametric*, J. P. Florens *et al.* (Editor). Lecture notes in Statistics, **16**, 13–35, Springer Verlag, New York.

76. Titterington, D. M., Smith, A. F. M. and Markov, V. E. (1986): *Statistical Analysis of Finite Mixtures of Distributions*, John Wiley and Sons, New York.

77. Tukey, J. W. (1960): *A survey of sampling from contaminated distributions, Contributions to Probability and Statistics*, I. Olkin *et al.* (Editor), Stanford University Press.

78. Veale, J. R. and Huntsberger, D. V. (1969): Estimation of mean when one observation may be spurious, *Technometrics*, **11**, 331–339.

79. Veale, J. R. (1972): Inference procedures for gamma populations when spurious observations may be present, Ph.D. Thesis Iowa State University, Ames, Iowa.

80. ——— (1975): Improved estimation of expected life when one identified spurious observation may be present, *Jour. Amer. Stat. Assoc.*, **70**, 398–401.

7

Some New Approaches to Problems in Storage and General Reliability Systems

N. Venugopal

Dept. of Statistics, S. V. University
Tirupati-517502, India

7.1 INTRODUCTION

The indispensibility of the tools and techniques in the theory of stochastic processes from the applicational viewpoint and the consequent operational value is now a well recognised and widely accepted fact. It is hardly an overstatement if one asserts that any serious piece of research work being reported in current scientific investigations invariably employs some kind of random processes or the other. This is primarily so, because if one wishes to come to grips with real life problems of practical interest, one has to deal with probabilistic systems that are dynamical.

The basic problem, therefore, is one of comprehending the 'dynamical probabilistic structures' governing the real life random phenomena that ever evolves. We have thus to deal with some basic probability space (Ω, B, P) and an abstract index set (deterministic and progressive), T, over which, a random variable that is defined on the basic probability space, keeps evolving: we thus have a typical random variable (r.v.) with two arguments: w and t, $w \in \Omega$ and $t \in T$. Let us denote this by $X(w, t)$.

The collection $\{X(w, t)\}$ is a typical random process. There, of course, are the two typical ways of looking at $\{X(w, t)\}$: (i) a family or train of r.v.'s $\{X(\cdot, t), t \in T\}$, and (ii) a set of random functions $\{X(w, \cdot), w \in \Omega\}$. The trivial case of $X(w, t)$, for fixed w and fixed t corresponds to the usual r.v. in the classical (static) probability theory.

Usually, we suppress 'w' for the sake of brevity and denote by $\{X(t)\}$, $t \in \mathrm{T}$ (that is the ensemble or collection of sample paths or trajectories over $t \in T$) a stochastic or random process. The probabilistic (dynamical) structure governing such a process is comprehended through the family of joint cumulative probability distribution functions $\{F_n(x_1, x_2, \ldots, x_n; t_1, t_2, \ldots, t_n)\}$, $n \in \mathbb{N}^+$, obeying the two conditions of (Kolmogorov's) consistency. Again, random processes are classified into four convenient types according to the continuous or discrete nature of $\mathbb{X}$ and T sets, where $\mathbb{X}$ is the state space defined by: $\mathbb{X} = \underset{t \in T}{U} S_t$,

$$X(t) : \xrightarrow{X(t)} S_t$$

It is now clear that a stochastic process is well defined in terms of $\mathbb{X}$, T and $\{F_n(\ldots;\ldots)\}$.

After quickly reviewing the preliminaries in the theory of stochastic processes, we make the important observation that comprehending the probabilistic structures governing the process of interest in terms of F_n's (in order to extract answers or solutions to practical problems) is in many situations cumbersome and at times intractable.

A review of applications of stochastic processes would testify to the truth of the above statement (see for example, Bartlett [12]).

It is in this context that point processes play a prominent role in terms of demonstrating their special and particular abilities to tackle a large class of problems of practical interest concerning 'highly localised events', with elegance, facility and within a sort of 'unified framework'. The number and diverse class of physical phenomena for which this type of stochastic (point) process provides a reasonable model is surprisingly large! Excellent accounts of the varied applicational scope of point processes are reported in Lewis [13], Donald and Synder [7] and Iosifescu and Tautu [11].

7.1.1 Point Processes

Point processes are prominently used in a variety of areas ranging from communication theory (laser transmission, photon detection), operations research (counter models, computer net-works) to nuclear medicine (radio-active tracers) and neurophysiology (spike trains along nervous fibres). The dynamical stochastic models that are proposed are broadly based on the two primary approaches, namely, (i) through the so-called 'stochastic intensity', and (ii) through 'martingale theory' (see, for details, Bremaud [5]).

In what follows, we carry out the exposition in some depth and detail and dealing with a ramification of the former approach. First, we observe that a point process over the half line $(0, \infty)$, can be characterized in any one of the following three equivalent ways:

(i) as a sequence of non-negative random variables (interval process of inter-point spacings),

(ii) as a discrete random measure, and

(iii) through its associated counting process (counting statistics).

The basic results leading to the one-to-one measure-preserving transformation of a general point process $(\mathbb{X}, B, P)$ into a 'counting process' $(\mathbb{n}, \mathbb{B}_n, \mathbf{P}_n)$ are established (in Moyal [15]). The pioneering suggestion of highlighting the applicational potentialities of point processes for storage models (in particular, the queueing systems) is due to Kendall [12]. An able follow up of the suggestion was carried out by Srinivasan [24] resulting in some notable contributions and later by Venugopal and others (see [25]-[31].

Definition 1. A random point process is a stochastic model for a physical phenomenon characterized by highly localised events distributed as a random stream of successive occurrences of point-events, the point events being understood as the occurrence of a phenomenon at an idealised time-point in the time continuum (domain), T, where the ensemble of the phenomenon of the point process is defined as a collection of discrete set of points belonging to T.

We remark here, that an observer of point-phenomena is often not directly concerned or interested in the point realisations themselves but rather in the underlying mechanism that influences or produces the occurrences of points and the inference that may be drawn about the mechanism.

Such an understanding of the underlying mechanism is achieved through the comprehension of the probabilistic structure governing the ensemble. This is accomplished in the two equivalent ways:

(i) through the time-continuous counting process, $N(t)$, $t \in T$, associated with the point process, $N(t)$ denoting the r.v. depicting the number of point events at epoch τ such that $\tau \in [t_0, t]$, generally, t_0 is chosen as origin, and

(ii) the interval processes:

$$X(n) = t_{n+1} - t_n,$$
$$X(0) = t_1, n = 1, 2, \ldots, \tag{1}$$

where t_n's are the epochs of point event occurrences.

That is, the probabilistic structure governing the associated counting and interval processes also govern the corresponding point process and this perspective proves thoroughly adequate. The 'popular' assumptions on $N(t)$ are:

A_1 : All epochs are distinct,

A_2 : $N(t) < \infty$ for finite intervals,

A_3 : $N(t) = \infty$ for infinite intervals, and

A_4 : Occurrence of point-events is stochastic in nature, that is, these do not manifest at predetermined epochs. (2)

7.1.2. The Product Density Approach

Rice (1944) was the first to use a type of characterization to study the zero-crossings of a continuous random process and later this approach was developed by Ramakrishnan (1950) which has come to be known as the "product density approach" (see also Bhabha [3]). We shall now discuss the approach and its power in operational use.

Let $P(n, x, t)$ denote the probability that 'n' random point-events characterized by a parametric value in $(x, x + dx)$ manifest at the epoch 't'. We now define $f_1(x, t)$ as:

Definition 2.

$$P(1, x, t) = f_1(x, t)\, dx + o(dx),$$
$$P(o, x, t) = 1 - f_1(x, t)\, dx + o(dx),$$
$$P(n, x, t) = o(dx),\ 2 \leqslant n \in \mathbb{N}^+, \tag{3}$$

so that if $dN(t, x)$ represents the number of entities in the state $(x, x + dx)$, at epoch, t, then:

$$\begin{aligned} E\{dN(t, x)\}^m &= E(n^m) \\ &= f_1(x, t)\, dx + o(dx),\ m \in \mathbb{N}^+. \end{aligned} \tag{4}$$

The result (4) has the interesting property that all the moments of the r.v., 'n' are equal and also are equal to the probability that 'n' takes the value 'unity', thus exhibiting the Poissonian character.

Further, we observe that:

$$\begin{aligned} E[N(t, b) - N(t, a)] &= E \int_a^b dN(t, x) \\ &= \int_a^b E[dN(t, x)] \\ &= \int_a^b f_1(t, x)\, dx, \end{aligned} \tag{5}$$

where $N(t, x)$ denotes the r.v. depicting the number of entities in $(0, x)$ at epoch, 't', $x = a, b\ (>a)$.

The result (5) thus makes it manifestly clear that while $f_1(t, x)$ has a probability density interpretation, its integral over x in (a, b) yields the mean (average) number of points (or entities) in the range (see also Bartlett [2]). One maÿ recall here the interesting analogy with the result that the expectation (average) of an indicator function of a r.v. is also the probability of the set when the indicator has the count '1' or assumes the value '1'. In fact, $f_1(x, t)$ is called the 'product density' (of order 'one'), for this reason (on the product space $\mathbb{X}$, $x \in T$). Definitions and other results concerning higher order product densities (needed in connection with second (variance) and higher (moments) are given in Srinivasan [24]. An alternative and interesting account of product densities in terms of the intensity process $\{\lambda_t, t \geqslant t_0\}$ of a self-exciting point process is given in Donald L. Synder [7]. We now observe that these product densities completely characterize a point process.

Usual investigations in the literature start with looking at $\{X(t)\}$ as a general random process; thus problems arise in the manipulative setup for deriving explicit results depending on the context, necessitating varied methods to tackle the situation (see [6] and [20]). However, if we look at the general $\{X(t)\}$ process as a point-event (counting process of

'x-level' crossings from above or below (Srinivasan [24]), the entire complexion of the attack on a large class of stochastic models substantially changes. This sort of perspective at once facilitates entry of product densities into the manipulative set-up resulting in considerable simplicity with the cognate virtues of directness, and elegance in approach to problem solution. Below we elaborate this idea and method through an illustration.

Looking at the general $\{X(t)\}$ process as a point process of x-crossings (from above or below), we introduce typical product densities of the type:

$$P(x, y, t) \stackrel{\text{def}}{=} \underset{\Delta \to 0^+}{\text{Lt}} \; Pr\{0 < x < X(t) \leqslant x + \Delta,\; I_t,\, E_t \mid I_0\}/\Delta, \tag{6}$$

where I_t, E_t and I_0 respectively denote the intermediatory conditions in $(0, t)$, and end condition at 't' and initial conditions at origin, i.e., $t = 0$.

By cleverly choosing these conditions, depending on the model and context, we observe that a large class of practical models could be solved.

Venugopal *et al.* ([25]–[31]) considered a dual type of a general storage model GI/GI/1/∞/FIFO as follows: Let $\{X(t)\}$ process describe the storage level as a non-negative continuous r.v. evolving with time (t) and is exposed to two constituent processes (assumed to be two independent renewal processes), the output, governed by $g(\cdot)$ and the inter-output time process, governed by $h(\cdot)$ where the input increases time-deterministically at a unit rate while the release (output) is stochastic and instantaneously depletes the $X(t)$ level. The basic change in this type dual model may be noted thus to consist in trading the stochastic and deterministic natures of input and releases (output).

For this model they introduced the product density, $M(x, y, t)$ where $M(x, y, t)$ is defined as in (6) with

$I_t = X(u) > 0,\ u \in [0, t]$

$E_t = X(t_+) < X(t)$, t being an epoch of point-event (release) occurrence

and

$$I_0 = X(0_-) > X(0) = y. \tag{7}$$

To solve for $M(x, y, t)$, the following integral equation involving Schwarzian density (measure), $\delta(\cdot)$ was formulated and established:

$$M(x, y, t) = U(x - y)\delta(x - y - t)h(t) + \int_0^t h(u)\, du \int_0^{y+u} M(x, y + u - v, t - u) g(v)\, dv. \tag{8}$$

Recall that $\delta(\cdot)$ is the Schwarzian measure accounting for the 'impulse' at 't' in the absence of demand (pull) in $[0, t]$, while $U(\cdot)$ is the Heaviside-unit-step functions.

Applying the L.T. technique to (8) and using simple arguments for 'analytic functions' in complex variable theory, we obtain the double

L.T., $M^*(x, s, p)$ of $M(x, y, t)$, with respect to the arguments y and t (for the special case of $M/G/1$) as:

$$M^*(x, s, p) = \frac{\lambda[e^{-sx} - e^{-(\lambda+p)x} - G(\lambda + p)M^*(x, \lambda + p, p)]}{\lambda + p - s - \lambda G(s)}, \qquad (9)$$

where $G(\cdot)$ is the L.T. of $g(\cdot)$ and $M^*(x, \lambda + p, p)$ is the constant to be evaluated for obtaining the explicit result (and lift the 'Laplacian curtain'!). This is accomplished by employing integral representation technique (due to Venugopal [25]).

7.1.3. The Technique

Appealing to the analytic property of $M^*(x, s, p)$ in the right half plane ($s > 0$), the constant $M^*(x, \lambda + p, p)$ can be evaluated by arguing that numerator (Nr) of (9) vanishes at $s = \delta$, δ being the zero with positive real part (R.P.) of the denominator (Dr) of (9) which leads to:

$$M^*(X, \lambda + p, p) = \frac{e^{-\delta x} - e^{-(\lambda+p)x}}{G(\lambda + p)}. \qquad (10)$$

The following result is now proved:

Lemma. $e^{-\delta x}$ is uniquely determined by the integral representation:

$$e^{-\delta x} = \underset{s\to\delta}{\text{Lt}} \frac{\lambda + p - s - \lambda G(s)}{(s - \delta)2\pi i} \int\limits_{\substack{\phi - i\infty \\ \phi > 0}}^{\phi + i\infty} \frac{e^{-\eta x}}{\lambda + p - \eta - \lambda G(\eta)}\, d\eta. \qquad (11)$$

Using, (10) and (11) in (9), we have

$$\widetilde{M}(x, y, p) \overset{\text{def}}{=} \int_0^\infty e^{-pt} M(x, y, t)\, dt$$

$$= \frac{1}{2\pi i} \int\limits_{\substack{\tau - i\infty \\ \tau > \text{Re.}\,\delta(>0)}}^{\tau + i\infty} \frac{e^{sy} M^*(x, s, p)}{\lambda + p - s - \lambda G(s)}\, ds$$

$$= \frac{\lambda}{2\pi i} \int\limits_{\tau - i\infty}^{\tau + i\infty} \frac{e^{-s(x-y)}}{\lambda + p - s - \lambda G(s)}\, ds$$

$$- \frac{1}{2\pi i} \int\limits_{\tau - i\infty}^{\tau + i\infty} \underset{s\to\delta}{\text{Lt}} \frac{e^{sy}}{(s - \delta)}\, ds$$

$$\times \left\{ \frac{1}{2\pi i} \int\limits_{\phi - i\infty}^{\phi + i\infty} \frac{e^{-\eta x}}{\lambda + p - \eta - \lambda G(\eta)}\, d\eta \right\}$$

$$= \frac{\lambda}{2\pi i} \int\limits_{\substack{\sigma - i\infty \\ 0 < \sigma < \text{Re.}\,\delta}}^{\sigma + i\infty} \frac{e^{-s(x-y)}}{\lambda + p - s - \lambda G(s)}\, ds, \qquad (12)$$

obtained by shifting the line ($\tau - i\infty$ to $\tau + i\infty$) to the left of the line ($\sigma - i\infty$ to $\sigma + i\infty$) and then evaluating the respective line integrals. Thus $\widetilde{M}(x, y, p)$, the characteristic function (ch. fn.) of the random time to x-crossing at 't' from the initial level 'y' is explicitly obtained.

Using the pattern of manipulative framework explained above (with suitable ramifications and consequent manipulative work) Venugopal *et al.* reported many explicit results, for characteristics of interest in storage systems (see [25]–[29]).

7.1.4 The Unified Approach

Introducing a crucial modification into the modelling setup described above, Venugopal and Meenakshi [30] further brought out the potentials of the product density approach in terms of developing unified method whereby results important in reliability contexts (e.g., explicit results for Mean Time Before Failure (MTBF) for a variety of reliability systems) can be extracted using the same manipulative techniques that are applicable to storage systems.

The distinguishing feature of the method further consists in the fact that the results are obtained for MTBF corresponding to different 'operating levels' at which the system is observed to be failure-free, as distinct from the classical work in reliability which limits its view-point to only 'up' states without reference to the "functioning efficacy" at the observed 'up' state.

The crucial modification introduced into the model proposed by Venugopal *et al.* [27], [30], [31] is that: "the 'random depletion' at an epoch (t) is stipulated to be sharp enough to reduce the storage level to an extent lower than that at the just previous epoch of release." The implication of this condition is that E_t in (7) is modified as follows:

$$E_t = X(t+) < X(t) \text{ and } X(t'+) < X(t''+) \ \forall\ (t'', t') \text{ in } (0, t) \text{ such that } t'' < t'. \tag{13}$$

Consequent upon this modification the interpretation of $X(t+)$ and the other cognate concepts is carried on as follows: The storage level $X(t+)$ is interpreted as 'operating level' (or 'stress level' the system can withstand or the 'residual life' of the system). Thus the 'positive' way of looking earlier at the storage level $X(t+)$ is now changed to a 'negative way' (for more details see [29]–[31]). For comprehending the random time to failure (from an epoch of continuous failure-free operation), the periods of 'idleness' do not matter (in view of [13]) and thus the modelling set up used for storage system now proves adequate.

Now connecting $f(\cdot)$ and $g(\cdot)$ in (8) to laws governing the operating and quiescent periods and the system subjected to constant hazard rates, which are related to $g(\cdot)$ in terms of the poles of $G(\cdot)$, Venugopal and Meenakshi [31] obtained the following expression to MTBF, $E(\tau)$, for a general G/G/1 type of system as:

$$E(\tau) = -G'(0) \Big/ \Big\{ l + \sum_{j=1}^{m} H(-z_j)\bar{G}(z_j)/z_j \Big\}, \tag{14}$$

where $H(\cdot)$ is the L.T. of $h(\cdot)$ and $\bar{G}(\cdot)$'s are the residues of $G(\cdot)$ at its simple poles z_j's, $j = 1, \ldots, m$ (the extension to the case of multiplicity of poles is straightforward).

Specialising to the sase of M/M/1 system with parameters λ and μ respectively, we obtain the result:

$$E(\tau) = (\lambda + \mu)/\mu^2, \tag{15}$$

agreeing with that in Gaver [9].

Several ramifications and interesting interpretations of the results derived on the basis of unified approach are given in Venugopal ([29]–[31]), the interpretations connecting the contexts in queueing, dam and reliability models.

7.1.5 Duality Principle

We now observe that the 'duality' relations in this context emanate from the fact that if $X(t)$ and $Y(t)$ are storage levels in the two storage models (for our model as its dual as given in earlier literature) for the finite capacity situation, $k(< \infty)$, then these two random variables are connected by the dual relationship:

$$X(t) \equiv k - Y(t)$$

or

$$Y(t) \equiv k - X(t).$$

Now noticing a "reflection principle", similar to the one given in Billinglley [4], as distinct from the reflection principle given in Feller [8] and others, we have developed duality methods to extract results for one type of model from its dual and vice versa with some simple changes in manipulative details.

Some instances of applicational use of the dual approach are also cited, though there still exists a large scope for further work (see also Prabhu [20] and Cohen [6]).

7.2 OPTIMAL RELIABILITY SYSTEMS

Of late keen interest is being evinced in regard to optimisation methods based on varied viewpoints for a variety of reliability systems. We may broadly classify these viewpoints as follows:

(i) MTBF (mean time before failure) performances of general reliability systems; alternative terminology used in literature for MTBF are ATF (Average Time to Failure) and MTTF (Mean Time to Failure).

(ii) Optimal number of units (N^*) comprising a parallel and other complex reliability system.

(iii) Optimal replacement or repair-stage (n^*), and

(iv) Optimal replacement times (T^*).

Consequently, the optimality criteria are based on either increased MTBF performances (see for example Rade [22]) or on minimising certain cost functions chosen for the purpose (see for example, Nakagawa [16]–[19]).

It can be said that most works are based on the assumption of "IDENTICAL UNITS" in a (parallel) system and somewhat restricted types of cost functions. There is scope for more realistic and operationally useful work, when one relaxes, (a) the condition of "identical units", and (b) introduces more general cost functions (for example, covering salvage costs, maintenance costs and so on). To minimise these certain optimisation techniques may be developed. There is the other perspection for developing cost-optimal (N^*, n^*) parallel systems. Again, there exists further scope when one extends this 'cost-optimal' concept to more complex reliability systems such as parallel-series and series-parallel systems. In fact, these extensions lead us to the so-called 'space-optimal' systems, an important class of complex reliability models.

In this section, we report the recent results of Venugopal *et al.* ([32]–[44]) in this direction covering: cost, repair and space-optimal reliability systems ('terms' being introduced by us).

7.2.1 An Overview

The earliest work in optimisation methods may be traced to the pioneering work due to Lokta [14]. He considered a model in which the equipment is subjected to random failures (but the state of equipment is always known with certainty) and developed a policy of replacement after the equipment's failure.

Barlow and Proschan [1] investigated planned replacement policy for a unit based on replacement and exchange cost. They proposed the following three replacement policies: (i) strictly periodic replacement, (ii) random periodic replacement, and (iii) sequentially determined replacement.

Rade [29] considered 'N' identical unit parallel systems subjected to shocks, generated through a random environment according to a "renewal point process" governed by a known law. He assumed that shocks cause the components to fail independently of each other with the same probability (known). For this system results for MTBF are developed. Later Nakagawa [17] developed replacement policy for Rade's model by incorporating acquisition and exchange cost considerations.

Nakagawa [18] considered a parallel system which consists of N identical units and developed optimal N^* by minimising the cost function $C(N)$ given by:

$$C(N) = \frac{NC_1 + C_2}{\mu_N},$$

where C_1 = acquisition cost of each unit

C_2 = additional cost replacement for a failed system, and

μ_N = MTBF of the systems.

He further obtained optimal replacement time-epochs (T^*) and the jointly optimal pair (N^*, T^*).

He further developed a procedure (see [19]) leading to optimal n^* for a K-out-of-n systems when the failure rate of each element is constant.

A good account of preliminaries and other earlier developments are well reported in Gnedenko *et al.* [10].

7.2.2 Motivation for Our Work

We may now summarise the general content of the investigations and results on optimisation methods for reliability systems as emerged in the above overview as follows:

(i) Basically, optimal number of 'identical' units (N^*) comprising a reliability system and the optimal replacement times (T^*) are sought to be obtained, for simple, parallel as well as more complex reliability systems such as K-out-of-n systems, through techniques of minimising certain cost functions based on rather restricted type of cost functions (for example, only acquisition and exchange costs are taken into consideration).

(ii) Optimal reliability systems (case of identical units) are sought to be developed based on a desired level of MTBF performance, sometimes linking this performance, with (limited) cost considerations.

We thus notice the following new lines for further work: (a) Cost functions could be structured to be more realistic and hence more general by bringing in a variety of other operationally relevant cost factors like salvage costs, maintenance costs and per-unit repair costs (when the additional feature of 'repair-sanction' is provided). That is, the purpose is to introduce more general cost functions minimising which optimal systems are developed with added advantage.

Further, we observe that the criterion of optimal repair stage (n^*) is operationally more convenient in terms of practical considerations than the concept of optimal replacement times (T^*). Thus optimal techniques based on optimal repair stages (n^*) are perhaps more desirable especially for the situation obtaining in developing countries like India.

There is another dimension to the optimisation techniques, when these are to be developed in the face of evolution of a system through technological progress. Optimisation methods, taking into account technological progress, is thus an exciting and promising area for work.

In addition, optimisation techniques for more complex systems like series-parallel and parallel-series systems (which are so relevant to mass communication and aero-space systems) are needed in no small measure.

(b) Optimal systems related to increased MTBF performances are based on the assumption of identical units constituting a parallel system. We observe that relaxing this assumption of 'identical units' has several operational and practical advantages. The results for general systems (comprised of not necessarily identical units) have large relevance in the present days of uncertain and fluctuating market conditions. Motivated by these factors, we present below a brief survey of our results reported in [32] through [44].

7.2.3 Cost-Optimal Systems

(a) Cost-optimal systems with maintenance costs

We now describe procedures for obtaining (i) the optimal N^* for a prefixed repair stage 'n', (ii) the optimal n^* for a given 'N', and (iii) the joint optimal pair (N^*, n^*) for a N-identical unit parallel system for the following model. The optimality criteria is based on minimal expected costs per unit time of operation.

The Model: Cosider a parallel system comprising of 'N' identical units with the two states: operating or failed. The system is repaired or replaced only after the system's failure. The system is assumed to be as good as new after each repair, that is, its MTBF performance remains the same. The maintenance cost increases with 'N' and further, the repair times increase with the number of failures. The time span is assumed to be infinite.

Let

$C_1 e^{-a/N}$ = the maintenance cost for 'N' unit parallel systems during $\mu(N)$ units of time, the constant 'a' (>0) characterizes the decay parameter. The cost structure is proposed in accordance with the physical experience.

Thus the expected cost per unit of time, in the steady states, is given by:

$$C(N, n) = \frac{NC_1 + (n+1)C_1 e^{-a/N} + C_2 \sum_{i=1}^{n} V_i}{(n+1)\mu(N)}, \tag{16}$$

where V_i is the mean repair time for ith failure.

Noticing that $C(N, n)$ is a step function in 'N' as well as 'n', we obtain N^* (for fixed 'n') by minimising $C(N, n)$ using the pair of inequalities $C(N+1, n) > C(N, n)$ and $C(N, n) < C(N-1, n)$.

This technique of minimisation leads to the following pair of inequalities:

$$L(N) > C_2/C_1 \quad \text{and} \quad L(N-1) < C_2/C_1, \tag{17}$$

where,

$$L(N) = \frac{[(N+1) + (n+1)e^{-a/N+1}]\mu(N) - [N + (n+1)e^{-a/N}]\mu(N+1)}{\sum_{i=1}^{n} V_i[\mu(N+1) - \mu(N)]}. \tag{18}$$

Thus, the N satisfying the inequalities in (17) turns out to be optimum.

Similarly, the optimum n^* (for fixed 'N') is the one that minimises

$C(N, n)$, i.e., the one satisfying $C(n+1, N) > C(n, N)$ and $C(n, N) < C(n-1, N)$ or equivalently:

$$L(n) > C_1/C_2 \quad \text{and} \quad L(n-1) < C_1/C_2 \tag{19}$$

where,

$$L(n) = \frac{(n+1) \sum_{i=1}^{n+1} V_i - (n+2) \sum_{i=1}^{n} V_i}{N}. \tag{20}$$

It is established that $L(N)$ in (18) and $L(n)$ in (20) are strictly increasing in their respective arguments (N and n) and further tend to $+\infty$ as N or n (as the case is) tends to $+\infty$, so that these [$L(N)$ or $L(n)$] cross the finite value C_2/C_1 or C_1/C_2 just once. Thus N^* and n^* are each finite and unique (for details see [31] and [32]).

Now using (17) and (19), we develop the following algorithm to obtain the jointly optimum pair (N^*, n^*).

The algorithm

The following steps constitute the algorithm:

(i) Set $N_1 = 2$ and find n_1^* using (19).

(ii) Set $n = n_1^*$ in (i) and find N_2^*, if exists, using (17).

(iii) Continue until $N_i^* = N_{i+1}^*$, where 'i' represents the ith iteration and at this stage the (N^*, n^*) is the optimum pair.

(iv) If in (ii), N_2^* does not exist, then set $N_1 = 3$ in (i); if again N_2^* (for $N_1 = 3$) does not exist, set $N_1 = 4$ in (i) and so on, and the steps may be continued as given in (i) through (iii) above.

Thus, when the iterations finally stop, the pair (N^*, n^*) at the ultimate stage represents the optimum pair.

We remark that while there is no mathematical proof ensuring termination of the iterative procedure, in practice, however, the procedure works out quite satisfactorily, in that the termination stage often is reached sufficiently early.

By way of illustration, we give the following empirical results to demonstrate the applicational use of the theoretical work (for details, see [33]).

Numerical example

For the case of negative exponential laws governing both failure times and repair times (not necessarily with the same parameters), in the later case with parameter chosen as $40/(1.1)^i$, $a = 2$, we have the following computed values for C_1/C_2, N^*, n^* and the resulting cost $C(N^*, n^*)/C_1$.

The physical meaning of C_1/C_2 is that it represents the unit acquisition cost per unit of repair time cost implying that increase in C_1/C_2 values generally corresponds to costlier unit. Tabulated values such as in Table 7.1 thus throw light on numerous alternative possibilities from varied viewpoints for the decision maker to choose that one which is most

Table 7.1. C_1/C_2 and (N^*, n^*)

C_1/C_2	N^*	n^*	$C(N^*, n^*)/C_1$
0.006	7	2	2.4319
0.010	4	2	1.8597
0.015	8	6	1.4494
0.030	6	8	1.0399
0.040	3	6	0.9288
0.050	4	8	0.8412
0.070	4	10	0.7401

acceptable. Further, work on the lines as indicated above with more alternatives for the concerned parametric values (if necessary with computer use), for more complex class of random laws (such as the Weibull law) governing the characteristics of the model, may prove immensely useful the operational and policy making purposes.

(b) Optional repair-stage with salvage costs

We now increase the scope in the last model by bringing in additionally the salvage cost considerations. Salvage costs are assumed to decrease with the increasing number of failures. Notice that the earlier costs of maintenance (in the previous model) are now replaced with the salvage costs.

Now, let:

$NC_1e^{-n/a}$ = Salvage value of the N-unit parallel systems after 'n' failures, where the parameter 'a' > 0 represents the decay pattern of salvage costs with increase in 'n'.

Thus, the expected cost per unit of time, in the steady state, is given by:

$$C(n) = \frac{NC_1(1 - e^{-n/a}) + C_2 \sum_{i=1}^{n} V_i(N)}{\sum_{i=1}^{n} \mu_{i+1}(N)}. \tag{21}$$

To obtain the optimal repair-stage n^*, we minimise $C(n)$, following the same lines as in (a) above. In this case $L(n)$ is given by:

$$L(n) = \frac{1}{N}\left\{\frac{\sum_{i=1}^{n+1} V_i(N) \sum_{i=1}^{n} \mu_{i+1}(N) - \sum_{i=1}^{n} V_i(N) \sum_{i=1}^{n+1} \mu_{i+1}(N)}{(1 - e^{-n/a}) \sum_{i=1}^{n+1} \mu_{i+1}(N) - (1 - e^{(-n+1)/a} \sum_{i=1}^{n} \mu_{i+1}(N)}\right\} \tag{22}$$

when n^* is cbtained through the pair of inequalities: $L(n) > C_1/C_2$ and $L(n-1) < C_1/C_2$ (for empirical work and other details see [39]).

(c) Optimal repair-stage based on a 'New Measure'

For the modelling set-up in (b), but without salvage cost considerations, we now obtain optimal repair-stage with the help of a New Measure (introduced for the purpose) in terms of absolute difference of repair and replacement costs.

The proposed measure, $\mu(n)$ is as follows:

$$\mu(n) = \left|\frac{NC_1}{\mu_1(N)} - \frac{C_2V_n(N)}{\mu_{n+1}(N)}\right|, \tag{23}$$

where, $\dfrac{NC_1}{\mu_1(N)}$ = the expected cost per unit of time (for the steady-state case) if replacement of the system is performed, and

$\dfrac{C_2V_n(N)}{\mu_{n+1}(N)}$ = the expected cost per unit of time for the steady-state case if repair is carried out.

Note that $C_2V_n(N)$ gives the expected repair cost after the nth repair.

We observe that the rationale underlying the proposed measure consists in the fact that the two expected costs (corresponding to repair and replacement) are sought to be brought as near to each other as possible in terms of the absolute difference between them.

The optimal repair-state n^* which minimises $\mu(n)$ is obtained through the pair of inequalities:

$$L(n) > C_1/C_2 \text{ and } L(n-1) < C_1/C_2,$$

where

$$L(n) = \frac{\mu_1(N)}{2N}\left[\frac{V_{n+1}(N)}{\mu_{n+2}(N)} + \frac{V_n(N)}{\mu_{n+1}(N)}\right]. \tag{24}$$

This approach is operationally quite helpful (for empirical work and other details see [40]).

(d) Optimal replacements with technological progress

Adopting the idea in the measure proposed in [23], in terms of consideration of absolute difference between the costs of interest, we now develop the optimal replacement policy for a single unit system with the important ramification, namely the presence of progress through technological evolution.

The Model: Let us consider a single unit system assumed to deteriorate with the number of failures. Further, after each failure, both the costs of maintenance and repair increase. Suppose that a cost C_3 is the acquisition cost of the system. Now a new system is introduced into the market as a result of technological progress which is distinct from the old system already in the market. For the new system the cost of maintenance and repair is assumed to be smaller compared to that for the old system but the acquisition cost, say C_4 is greater than C_3. However, the per unit time efficiency in terms of the MTBF performance of the old and new system are assumed to be the same.

We now propose the measure, $C(n)$ given by:

$$C(n) = \left| \frac{C_3 - C_3 e^{-n/a} + C_3 e^{-b/n+1}}{\mu_{n+1}} - \frac{C_4 + C_4 e^{-c}}{\tau_1} \right|, \tag{25}$$

where, $C_3 e^{-b/n+1}$ = the cost of maintenance and repair for the old system after 'n' failures, $b > 0$ and constant.

$C_3 e^{-n/a}$ = the salvage value of the old system after 'n' failures, $a > 0$ and constant.

$C_4 e^{-C/n+1}$ = the cost of maintenance and repair for the new system after 'n' failures; $0 < b < c$.

τ_1 = MTBF of new system subjected to no failures (i.e. $n = 0$).

and, μ_n = MTBF of the old system after $(n - 1)$ failures.

The physical interpretation of the decay parameters: a and b and the motivation for the measure in [25] are to be understood as already indicated. Now to obtain n^* which minimises $C(n)$, we use the pair of inequalities:

$$L(n) > C_4/C_3 \text{ and } L(n - 1) < C_4/C_3, \tag{26}$$

where,

$$L(n) = \frac{\tau_1}{2(1 + e^{-c})} \left\{ \frac{1 - e^{-(n+1)/a} + e^{-b/n+2}}{\mu_{n+2}} + \frac{1 - e^{-n/a} + e^{-b/n+1}}{\mu_{n+1}} \right\}, \tag{27}$$

Other details and empirical work are given in [34].

(e) Optimal parallel-series systems

We developed an optimisation procedure leading to Cost-Space Optimal (N^*, m, n^*) parallel-series systems where the three entries N^*, m and n^* respectively denote, paralleling number, the number of units connected in series and the optimal repair-sanction accorded to the systems (that is the most economical number of repairs for a system to undergo before it is replaced).

For a parallel-series of order (N, m), we developed a procedure to obtain the optimal n^* for given 'N' and 'm', which leads to the 'cost optimal' (N, m, n^*) system. Next, we developed a procedure to obtain the optimal N^* for a given 'n' leading to the 'space optimal' (N^*, m, n) system. Now combining these two procedures we proposed an algorithm leading to the "cost and space optimal (N^*, m, n^*) system-design". This terminology is introduced by us (for more details see [37]).

(f) Optimal series-parallel systems

We developed an optimal procedure leading to the most economical number of parallel units (N^*) in a (m, n) series-parallel systems (see [36] for details).

(g) Optimal K-out-of-N systems with repair provision

Now for the other type of reliability system, namely the K-out-of-N systems, we may bring in repair and maintenance costs into the model formulation and minimising the cost functions, the most economical (optimal) number of units N^* in a K-out-of-N system may be obtained. The results are under reporting (see [43]).

7.2.4 Optimal MTBF—General Systems

So far, we have discussed cost-optimal systems basically developed through cost minimisation technique. We shall now explore the scope to evolve optimal systems based on their increased or some desired level of MTBF performance. We consider here general systems comprised of not necessarily identical units.

(a) General systems—Units failing independently

Let us consider a parallel system with 'n' components which are 'not identical' in the sense that these components, while failing independently of each other have different but known probabilities of failure, p_i, $i = 1, 2, \ldots, n$. The failures are assumed to occur as a result of 'shocks' generated by a 'random environment' in which the system operates. These shocks are assumed to be generated by a "renewal point process" whose interval properties are governed by a known law, $F(\cdot)$, with $F(0) = 0$.

The following theorem has important applicational consequences (see [41]).

Theorem 1. In the notation given earlier (see [41]),

$$\bar{H}_n^*(s) = \bar{F}^*(s) \sum_{i=1}^{n} \sum_{1} (-1)^{i+1} \Big/ \left\{1 - \prod_{k=l_1}^{l_i} q_k F^*(s)\right\}, \tag{28}$$

where the l_i's ran over the (n_i) possible combinations of the sub-systems.

As an important consequence of the theorem, we can develop a ratio for comparing a general (parallel) systems with an arbitrary basic component in terms of their MTBF-performances. This is given as

$$\frac{\mu_n}{\mu_1(p_1)} = \sum_{i=1}^{n} \sum_{1} (-1)^{i+1} p_1 \left\{1 - \prod_{k=l_1}^{l_i} q_k\right\}^{-1} \tag{29}$$

This result has important use from the operational viewpoint, for, the crucially important ratios $\mu_n/\mu_1(p_1)$ may be computed for a variety of p's, thereby making available a large variety of basic single components for comparison purposes (in terms of relative increased MTBF-performance with increase in n's).

In particular, if p_i's are all equal, say to 'p', then (29) neatly simplifies to

$$\frac{\mu_n}{\mu_1(p)} = \sum_{i=1}^{n} \binom{n}{i} (-1)^{i+1} \frac{1-q}{1-q^i}, \tag{30}$$

in agreement with the result in Rade [21].

(b) General systems—Dependent failures

We now relax even the rather simplistic assumption of independent failures of units in a general system and propose the following model with units failing according to a 'dependency pattern', explained in the following:

The units fail with different but known probabilities, $p_k(i)$'s, $p_k(i)$ denoting the ith units failure, probability when k units are operating. These $p_k(i)$'s are dependent in the sense that the $p_k(i)$'s form an increasing sequence as k's decrease. That is, $p_k(i) > p_{k+1}(i)$, for every 'i' and 'k'. These failures are consequent to the shocks generated by a random environment in which the system operates. A shock may or may not produce a component-failure or may produce failures of more than one component. The epochs of shock occurrence are assumed to form a renewal point process whose interval properties are governed by a known law $F(\cdot)$, with $F(0) = 0$.

The following theorem has important operational use (see [42] for other details and empirical work).

Theorem 2.

$$\bar{H}_n(t) = \bar{F}(t) + F(t) \prod_{i=1}^{n} q_n(i)\bar{H}_n(t)$$

$$+ F(t) \sum_{(k)} \prod_{i=1}^{k} q_k(i) \prod_{i=1}^{n-k} p_k(i)\bar{H}_k(t), \tag{31}$$

where (k), runs through the $\binom{n}{k}$, $k = 1, 2, \ldots, (n-1)$, varieties of distinct combinations resulting from the 'k' non-identical components out of the 'n' non-identical components, so that the total number of distinct terms in the sum is given by

$$\binom{n}{1} + \binom{n}{2} + \ldots + \binom{n}{n-1} = 2(2^{n-1} - 1).$$

We have now the explicit result for μ_n in terms of the known $F(\cdot)$ and $p_k(i)$'s, and is given by

$$\mu_n = V_n\left\{\mu + \sum_{(k)} \prod_{i=1}^{n} q_k(i) \prod_{i=1}^{n-k} p_n(i)\mu_k\right\}, \tag{32}$$

where

$$V_n = \left[1 - \mathop{\pi}_{i=1}^{n} q_n(i)\right]^{-1}.$$

Now, specialising to the case of parallel system with identical components, we observe that $\bar{H}_n(t)$ and μ_n reduce to:

$$\bar{H}_n(t) = \bar{F}(t) + F(t)q_n^n\bar{H}_n(t) + F(t)\sum_{k=1}^{n-1}\binom{n}{k}q_k^k p_k^{n-k}\bar{H}(t) \tag{33}$$

and

$$\mu_n = V_n\left\{\mu + \sum_{k=1}^{n-1}\binom{n}{k}q_k^k p_k^{n-k}\mu_k\right\}. \tag{34}$$

These results are in agreement with those in Rade [21].

(c) Optimal replacements—Independent failures

Based on the model in (a), we develop optimal replacement policy as follows:

Consider a parallel system which consists of 'n' non-identical components and the system fails when all components have failed. The system is situated in a random environment which generates shocks according to a general random law with mean $1/\lambda$.

These components, while failing independently of each other have different but known probabilities of failure, p_i, $i = 1, 2, \ldots, n$. We propose the replacement policy for the system as follows: The system is exchanged preventively before failure, if the total number of failed components is: $(k + 1), (k + 2), \ldots$ or $(n - 1)$ and it is replaced if all components fail. The system is left alone, if number of components failed $1 \leqslant k(\leqslant n)$. We now obtain the optimum k^* as that one which minimises the expected cost per unit of time (in the steady state case). The time span is assumed to be infinite.

7.2.5. Optimal Replacement Policy

Let Y_i be the total number of components which are failed by the ith shock $(i = 1, 2, \ldots)$. Then, the probability that the system is replaced after failure is given by:

$$A_1(k) = Pr(Y_1 = n) + \sum_{j=2}^{\infty}\sum_{r=0}^{k} Pr\{Y_1 + Y_2 + \ldots + Y_{j-1} = r \text{ and } Y_j = (n-r)\}$$

$$= \mathop{\pi}_{i=1}^{n} p_i + \mathop{\pi}_{j=2}^{\infty}\sum_{r=0}^{k}\left\{\left(\sum_{\binom{n}{i_1}}\mathop{\pi}_{i_1} p_k \mathop{\pi}_{n-i_1} q_{k_1}\right)\left(\sum_{\binom{n-i_1}{i_2}}\mathop{\pi}_{i_2} p_{k_2}\mathop{\pi}_{n-i_1-i_2} q_{k_2}\right)\cdots\right.$$

$$\left.\times\left(\sum_{\binom{n-i_1-i_2\ldots i_{j-2}}{i_{j-1}}}\mathop{\pi}_{i_{j-1}} p_{k_{j-1}}\mathop{\pi}_{(n-i_1-i_2\ldots i_{j-1})} q_{k_{j-1}}\right)\mathop{\Pi}_{n-r} p_{k_j}\right\}, \tag{35}$$

where $r = \sum_{i=1}^{j-1} i_1$, p_{k_j}'s and q_{k_j}'s represent the failure and operating probabilities of the units and the $\sum_{(.)}$ and $\pi_{(.)}$ run through the number of entries as indicated in the (.) below the summation (Σ) and products (π).

By a similar argument, the probability that the system is exchanged before failure is given by:

$$A_2(k) = Pr\{k+1 \leqslant Y_1 \leqslant (n-1)\} + \sum_{j=2}^{\infty} \sum_{r=0}^{k} Pr\{Y_1 + Y_2 + \ldots + Y_{j-1} = r \text{ and } (k+1-r) \leqslant Y_j \leqslant (n-1-r)\} \tag{36}$$

or equivalently $= 1 - A_1(k)$.

The mean time to replacement (MTR) of the system is given by:

$$MTR = \frac{1}{\lambda} \sum_{j=1}^{\infty} j \sum_{r=0}^{k} Pr\{Y_1 + \ldots + Y_{i-1} = r \text{ and } Y_i \geqslant (k+1-r)\}. \tag{37}$$

Now, the expected cost per unit of time, in the steady-state, is given by:

$$C(k) = \frac{C_1 A_1(k) + C_2 A_2(k)}{MTR}, \; k = 0, 1, 2, \ldots, (n-1), \tag{38}$$

where C_1 and C_2 respectively denote the replacement and exchange costs of the system. For a given 'n', the optimum k^* is now that 'k' which corresponds to the minimum expected cost among all the $C(k)$'s for $k = 0, 1, \ldots, (n-1)$.

Proof and other details are given in [35]. For the particular case of systems comprising of identical units, the results in Nakagawa [17] can be recovered from [35].

For the purpose of empirical illustration, we consider a 4 unit general system and compute the values $C(0)$, $C(1)$, $C(2)$, $C(3)$. Now the optimum k^* is obtained corresponding to the minimum value in the above $C(k)$'s, $k = 0, 1, 2$ and 3. Choosing some typical values for p_k's, we present the optimal k^* corresponding to some selected cases.

Table 7.2

n	p_1	p_2	p_3	p_4	k^*
4	0.20	0.20	0.20	0.20	2
4	0.30	0.30	0.30	0.30	1
4	0.40	0.40	0.40	0.40	1
4	0.50	0.50	0.50	0.50	1
4	0.60	0.60	0.60	0.60	1
4	0.20	0.30	0.40	0.50	1
4	0.30	0.40	0.50	0.60	1

Comments: We now observe that no great advantage is achieved by constituting a parallel system with the same quality (that is identical units) components in so far as the optimum replacement policy (k^*) is concerned. Further, as the above empirical work depicts, one can perhaps accomplish some economy in designing a system by going in for even slightly cheaper (in terms of units with larger p_k's) items.

These and other such operational conveniences may emerge by investigating general systems rather than systems with identical units. A basic purpose of the work is to highlight such applicational potentialities.

The optimal replacement policies for the model discussed above can be extended further to cover the case of components failing according to some 'stochastic dependency pattern'. For example, the failure probabilities may be taken to depend upon the number of shocks the system receives. These and other interesting consequences illustrated with empirical illustrative work are under reporting (see [43]).

REFERENCES

1. Barlow, R. E. and Proschan, F. (1982): Planned Replacement in K. J. Arrow, S. Karlin and W. L. Smith (Eds), *Studies in Allied Probability and Management Science*, Stanford University Press, Stanford, pp. 63–87.
2. Bartlett, M. S. (1966): *An introduction to stochastic processes with special reference to methods and applications*, 2nd edition, Camb. Univ. Press, London.
3. Bhabha, H. J. (1950): On the stochastic theory of continuous parametric systems and its applications to electron-photon cascades, *Proc. Roy. Soc.*, **A, 202,** pp. 301–332.
4. Billingsley, P. (1968): *Convergence of probability measures*, John Wiley, N.Y.
5. Bremand, P. (1981): *Point processes and Queues, Martingale Dynamics*, Springs-Valley, New York.
6. Cohen, J. W. (1969): *The single server queue*, North Holland Pub., Amsterdam.
7. Donald, L. Synder (1975): *Random point processes*, A Wiley Interscience Pub., John Wiley, N.Y.
8. Feller, W. (1963): *An introduction to Probability Theory and its applications*, Vol. II, Wiley Eastern, New Delhi.
9. Gaver, D. P. (1963): Random hazard in reliability problems, *Technometrics*, **5**.
10. Gnedenko et al (1969): *Mathematical methods of reliability theory*, Academic Press, New York.
11. Iositeseu, M. and Tautu, P. (1973): *Stochastic Processes and application in Biology and Medicine*, Springer Verlag, Berlin.
12. Kendall, D. G. (1951): Some problems in the theory of queues, *JRSS*, **B, 13,** pp. 151–185.
13. Lewis, P. A. W. (1972): *Stochastic point processes, statistical analysis, theory and applications* (Ed), John Wiley, N.Y.
14. Lotka, A. J. (1939): A contribution to the theory of self-renewing aggregated with special reference to Industrial Replacement, *Ann. Math. Stat.*, **Vol. 10,** pp. 1–25.

15. Moyal, J. E. (1962): The general theory of stochastic population processes, *Acta Math.*, **108**, pp. 1–31.

16. Nakagawa, T. (1978): The decision to replace a unit early or late in an age replacement problem, *Micro-electronics and Reliability*, **Vol. 19**, pp. 365–367.

17. ——— (1979) : Replacement problem of a parallel system in random environment, *J. Appl. Prob.*, **Vol. 16**, pp. 923-926.

18. ——— (1984): Optimal number of units for a parallel system, *J. Appl. Prob.*, **Vol. 21**, pp. 431–436.

19. ——— (1985): Optimization problems in *K*-out-of-*n* systems, *IEE Trans. Reliability*. **Vol. R–34**, pp. 248–250.

20. Prabhu, N. U. (1965) *Queues and inventories*, John Wiley, N.Y.

21. Ramakrishnan, A. (1950): Stochastic processes related to practices distributed in an infinity of states, *Proc. Camb. Phil. Soc.*, **46**, 595–602.

22. Rade, L. (1976): Reliability systems in random environment *J. Appl. Prob.*, **Vol. 13**, pp. 407–410.

23. Rice, S. O. (1944): Mathematical analysis of random noise, *Bell. Syst. Tech. Journal*, **23–24**, pp. 1–162.

24. Srinivasan. S. K. (1974): *Stochastic point processes and their applications*, Griffin, London.

25. Srinivasan, S. K. and Venugopalacharyulu, N. (1976): Analytic solution of some stochastic models of finite dam with bounded input, *Jour. of Math. & Phy. Sci.*, **Vol. 10, No. 2**, pp. 141–164.

26. Venugopalacharyulu, N. (1975): Analytic solution of some stochastic models of finite dam with bounded input or inter-input times, Ph.D. thesis (abstract), *Opsearch*, **Vol. 12, No. 3** and **4**, pp. 124–126.

27. ——— (1976): Finite dam—A multivariate analysis, Proc. of the Advanced Symposium on Probability and its applications, Annamalai Univ. Publication, pp. 457–471.

28. Venugopal, N. and Sarma, K. L. A. P. (1984): Explicit results for a finite banking model with a critical reserve level, Proc. of VIII ISTPA Conf. Delhi Univ., Wiley Eastern, pp. 100–104.

29. Venugopal, N. (1986): Some new approaches through product densities in storage and reliability systems, *The Journal of Orissa Mathematical Society*, **Vol. 5, No. 2**, pp. 163–175.

30. Venugopal, N. and Reddy, K. V. (1986): Operational efficiency of S.D. Models, Pro. Vol. I(A) of XXVIII National Convention of IIIE, NITIE, Bombay, pp. 272–277.

31. Venugopal, N. and Meenakshi Bai, M. (1987): Hazard models with stochastic over-times or Built-in Buffers, *IAPQR transactions*, **Vol. 12, No. 1**, pp. 27–39.

32. Venugopal, N, Rami Reddy, C. and Somasundaram, M. (1987): Cost optimal systems with fixed repair sanction, *Micro electronics & Rel.* (U.K.), **Vol. 27**, pp. 419–422.

33. Venugopal, N., Rami Reddy, C and Meenakshi Bai, M. (1988): Cost optimal (N^*, n^*) parallel reliability systems, *Intl. Journal of Systems Success* (U.K.), **Vol. 19, No. 8**, pp. 1623–1629.

34. Venugopal, N., Reddy, Y. K. and Rami Reddy, C. (1988): Optimal replacement policy for a single unit reliability model with technological progress, *Assam Statistical Review*, **Vol. 1, No. 2**, pp. 1–8.

35. Venugopal, N., Shaffi Ahmed, S. and Meenakshi Bai, M. (1988): Optimal replacement policy for general parallel reliability systems, *Intl. Journal of Systems Sciences* (U.K.), Vol. 19, No. 12, pp. 2605–2618

36. Venugopal, N. and Reddy, Y. K. (1989): Optimal (n, m, N^*) series-parallel systems with fixed repair facility, *Intl. Journal of Systems Sciences* (U.K.), Vol. 20, No. 2, pp. 229–233.

37. Venugopal, N., Reddy, Y. K. and Meenakshi Bai, M. (1989): Cost space optimal (N^*, m, n^*) parallel series systems, *Micro-electronics & Rel.* (U.K.), Vol. 29, No. 1, pp. 9–15.

38. Venugopal, N., Shaffi Ahmed, S. and Rami Reddy, C. (1989): Optimal repair-stage for K-out-of-N systems, *Micro-electronics & Rel.* (U.K.), Vol. 29, No. 1, pp. 17–19.

39. Venugopal, N., Reddy, V. K. and Rami Reddy, C. (1986): Cost-optimal replacement policy for n-unit parallel reliability system with provision for salvage costs, Proceedings of the VIII I.S.P.S. Conference, Kolhapur University, pp. 79–88.

40. Venugopal, N. and Rami Reddy, C. (1986): Optimal repair-stage policy for N-unit parallel reliability systems, Proceedings of the VIII ISPS Conference, Kolhapur University, pp. 71–78.

41. Venugopal, N. and Shaffi Ahmed, S. (1986a): Parallel systems with general components in random environments, *I. E. Journal*, **Vol. XV, No. 8**, pp. 7–10.

42. ——— (1986b): General parallel systems in random environment subjected to dependent stochastic failures, Proceedings of the International Conference of AMSE, **Vol. X**, pp. 48–60.

43. ———: Optimal replacement policy for general reliability systems with dependent failures (communicated).

44. Venugopal, N. and Krishna Reddy, Y.: Optimal K-out-of-N systems with repair provision (Communicated).

8

Random Power Series in One or Several Complex Variables, A Theorem of Steinhaus and Applications

J. Gopala Krishna and A. Satyanarayana Murty
Dept. of Math., Andhra University, Waltair, A.P., India

8.1 INTRODUCTION

Let, as usual, C^k denote the complex space, which is the cartesian product of k copies of the complex plane; and let

$$\sum_{n \in I} a_n z^n \tag{1.1}$$

be a power series (in z) over C^k, where $I = I^k$ denotes the set of all ordered k-tuples of non-negative integers z^n is written for $\prod_{j=1}^{k} z_j^{n_j}$ and the coefficients a_n are all complex valued random variables on the same Kolmogorov Probability Space $(\Omega, \mathcal{F}, P)$. (For the relevant background in Probability see Ref. [2], [4], [6]).

Extensive literature about (1.1) is available in the case of one variable (Ref. [11], [14] and [15]) and further when it reduces to a random polynomial (Ref. [3]). Much remains to be done in several variables. Some spadework has been done by the author in this regard (Ref. [7]) in the general case of k, when the coefficients a_n of (1.1) form a family of independent random variables which we shall assume throughout the rest of this work [see [5] for terminology and basic results].

8.2.1 Definition

A power series in C^k with (fixed) complex coefficients is said to be singular iff (if and only if) either the (open) domain of convergence of the series is not C^k, but is its natural domain (i.e., a domain beyond which the sum of the series cannot be continued analytically) or the series is entire (with C^k as its domain of convergence).

In the work cited above the author has established the essential non-randomness of the domain of convergence of (1.1) and posed the partially discussed problems connected with Borel conjecture concerning the a.s. singularity of (1.1) and the Blackwell conjecture concerning the resolution of (1.1) into a sum of an almost singular series and a power series with fixed coefficients (for the concerned background for $k = 1$ see Refs. [14] and [15]). The author has also discussed problems relating to (1.1) when it is an a.s. entire series.

As already cited above Borel (see [15]) expressed his view, when $k = 1$, that (1.1) is singular a.s. However the view-point needs a precise mathematical interpretation, by way of fixing a probability model, before it may be discussed. Ryll-Nardzewski [14] pointed out that the Borel conjecture is not true with reference to the Kolmogorov probability model. But Steinhaus [15] restricted himself to an independent random selection of the angles of the coefficients of (1.1) (with the moduli of coefficients fixed) when $k = 1$. His result may be formulated in terms of the Kolmogorov probability model as:

Theorem 1.1 Let

$$\sum_{n=0}^{+\infty} c_n z^n \ (z \in C)$$

be a complex power series with a positive finite radius of convergence, where, as ususal, C is the complex plane. Let $\{\phi_n, n = 0, 1, 2, \ldots\}$ be a sequence of real valued independent random variables on the Kolmogorov space $(\Omega, \mathcal{F}, P)$ such that each ϕ_n is uniformly distributed in $[0, 2\pi)$. Then the random power series

$$\sum_{n=0}^{+\infty} c_n \exp (i\phi_n(\cdot))z^n \ (z \in C)$$

has its circle of convergence as its natural boundary, a.s. or precisely, the set of all $\omega \in \Omega$ such that the series

$$\sum_{n=0}^{+\infty} c_n \exp (i\phi_n(\omega)z^n \ (z \in C)$$

has its circle of convergence as its natural boundary, is an event (i.e. a member of $\mathcal{F}$) and it has probability 1.

It is needless to say that Steinhaus' discussion precedes the establishment of a sound theory of Probability, in particular, that of Kolmogorov. It therefore requires to be streamlined.

In the present work, we establish, with the help of a number of recently established results, a generalisation of the above theorem in the case of one or more complex variables and offer a probabilistic proof of the fact that the domain of convergence of any (non-entire) power series in several complex variables is a domain of holomorphy.

8.2 PRELIMINARY RESULTS IN SEVERAL COMPLEX VARIABLES

We use the standard notation of [5] and [8]. Throughout this, k denotes an integer $\geqslant 1$ and as usual C^k denotes the cartesian product of k copies of the Complex plane, $C^1 = C$, $\mathbf{R}^k$ denotes the cartesian product of k copies of the real line $\mathbf{R}$ and $I = I^k$ denotes the set of all ordered k-tuples of non-negative integers. For any non-empty subset E of C^k, we write $|E|$ for the set

$$\{(|z_1|, \ldots, |z_k|) : (z_1, \ldots, z_k) \in E\}$$

and E^+ for the set

$$\{(z_1, \ldots, z_k) \in |E| : z_j > 0 \ (j = 1, 2, \ldots, k)\}.$$

We follow the convention of referring to $(z_1, \ldots, z_k)$, $(\alpha_1, \ldots, \alpha_k)$ and $(n_1, \ldots, n_k)$ etc. $\in C^k$ by their respective unsuffixed symbols z, α, n etc. For any $n \in I$, $\|n\|$ denotes $\sum_{j=1}^{k} n_j$, $n!$ denotes $\sum_{j=1}^{k} n_j!$. Further for any $z \in C^k$ and $n \in I$, z^n denotes $\prod_{j=1}^{k} z_j^{n_j}$ (where $z_j^{n_j}$ should be taken as 1 if $z_j = n_j = 0$ in the present discussion).

For any m, $n \in I$, we write $m \geqslant n$ or $n \leqslant m$ to mean $m_j \geqslant n_j$ $(j=1, 2, \ldots, k)$. For any $\alpha \in C^k$ and $r \in C^{k+}$, $P(\alpha, r)$ denotes the (open) polydisc (with centre at α and polyradius r) specified by

$$\{z \in C^k : |z_j - \alpha_j| < r_j \ (j = 1, 2, \ldots, k)\}$$

and $\bar{P}(\alpha, r)$ denotes the closure (in C^k) of $P(\alpha, r)$. Finally, for any $r \in C^{k+}$, P_r denotes the (open) polydisc $P(\bar{O}, r)$, where $\bar{O} = (0, \ldots, 0) \in C^k$ and T_r denotes the polycircle:

$$\{z \in C^k : |z_j| = r_j = (j = 1, \ldots, k)\}.$$

Throughout this work,

$$\sum_{n \in I} c_n z^n \quad (z \in C^k) \tag{2.1}$$

denotes a complex power series in C^k where c_n is a complex number for each $n \in I$.

In the case of one complex variable (when $k = 1$), the usual Hadamard formula (see [1]) characterizes the (open) domain of convergence of the power series (2.1) while in the case of several complex variables, a characterization of the (open) domain of convergence of the series (2.1) is given by

Theorem 2.1 (Lemma 1.2) of [7] or Theorem 3.7 of [5]). Let D be the domain of convergence of (2.1). Let ψ be any real valued function defined on $I = I^k$ such that

$$0 < \liminf_{\|n\| \to +\infty} \frac{\psi(n)}{\|n\|} < +\infty, \ 0 < \limsup_{\|n\| \to +\infty} \frac{\psi(n)}{\|n\|} < +\infty. \tag{2.2}$$

Let $\alpha \in C^k$ and let no component of α be 0. Then $\alpha \in D$ iff

$$\limsup_{\|n\| \to +\infty} |c_n \alpha^n|^{1/\psi(n)} < 1. \tag{2.3}$$

We now discuss about singular points of the power series (2.1). When

$k = 1$, it is well-known that the circle of convergence of (2.1) contains at least one singular point of the same, if its radius of convergence is positive and finite. Fuks provides an analogue of this result for $k = 2$ (see Theorem 2.6 of [5]). We need the extensions of these results for the general k (with proof also suitably modified to secure ease at presentation in the general case).

Theorem 2.1. Let D be the domain of convergence of (2.1) and let f be its sum function in D. Let $r = (r_1, \ldots, r_k) \in C^{k+}$ be a boundary point of $|D|$. Then there exists at least one singular point for (2.1) on the poly circle T_r.

It is convenient to have the following lemmas.

Lemma 2.1. Let $r \in C^{k+}$. Let U, V be open polydiscs with centres on T_r and let $U \cap V \neq \phi$. Then $P_r \cap U \cap V \neq \phi$.

Proof. When $k = 1$, the lemma reduces to a well-known property of circular discs in C. In the general case, let P_r, U and V be the cartesian products $\overset{k}{\underset{j=1}{X}} P_j$, $\overset{k}{\underset{j=1}{X}} U_j$ and $\overset{k}{\underset{j=1}{X}} V_j$ respectively, where P_j, U_j and V_j are open circular discs in C such that the centres of U_j and V_j lie on the boundary of P_j for $j = 1, 2, \ldots, k$. Now, since

$$P_r \cap U \cap V = \overset{k}{\underset{j=1}{X}} (P_j \cap U_j \cap V_j),$$

the lemma follows from its special case when $k = 1$.

Lemma 2.2. Let D, r, f be as in Theorem (2.1). Let S be a nonempty subset of T_r such that for each $\alpha \in S$, f admits an extension analytic at α. Then there exists an extension of f which is analytic at each point of S (i.e. there exists an open set U in C^k containing $P_r \cup S$ and a function g analytic in U such that $g = f$ on P_r).

Proof. By hypothesis, for each $\alpha \in S$, there exists a nonempty open polydisc Q_α with centre at α and a function g_α which is an extension of f and is analytic in $P_r \cup Q_\alpha$ (we recall that P_r is the polydisc $P(\bar{O}, t)$). Let $U = P_r \cup (\bigcup_{\alpha \in S} Q_\alpha)$. If $\alpha, \beta \in S$ and $Q_\alpha \cap Q_\beta \neq \phi$, then by Lemma (2.1), it follows that $P_r \cap Q_\alpha \cap Q_\beta \neq \phi$. Hence by the identity theorem concerning analytic functions of several complex variables (see, e.g. Proposition 1—Chapter 1 of [12]), it follows that $g_\alpha = f = g_\beta$ on $Q_\alpha \cap Q_\beta$. Hence there exists a function g on U such that

$$g(z) = \begin{cases} f(z) & \text{if } z \in P_r \\ g_\alpha(z) & \text{if } z \in Q_\alpha \text{ for some } \alpha \in S. \end{cases}$$

This g is obviously an analytic extension of f into U and hence the lemma.

Lemma 2.3. Let $r \in C^{k+}$ and let Δ be an open set in C^k-containing T_r.

Then there exists a $\delta \in C^{k+}$ such that $(r - 2\delta) \in C^{k+}$ and the polyannulus

$$A = \{z \in C^k : r_j - 2\delta_j < |z_j| < r_j + 2\delta_j \text{ for } j = 1, 2, \ldots, k\}$$

is contained in Δ.

Proof. Since T_r is compact, the distance d of $C^k - \Delta$ from T_r (with the Eucleadian norm of C^k) is > 0. Let us choose $\delta \in C^{k+}$ such that $(r - 2\delta) \in C^{k+}$ and $\|2\delta\| < d$.

Let $z \in A$. The proof will be complete if we show that dist $(z, T_r) < d$, since then z would be in Δ. Now let $z_j = |z_j| \exp(i\theta_j)$, where $\theta_j \in \mathbf{R}$ $(j = 1, 2, \ldots, k)$. Now the point ξ with $\xi_j = r_j \exp(i\theta_j)$ $(j = 1, 2, \ldots, k)$ is in T_r. The lemma now follows from the fact that dist $(z, T_r) \leqslant \|z - \xi\| < \|2\delta| < d$.

Proof of Theorem (2.1). Suppose, to get a contradiction, that every point of T_r is a regular point for the series (2.1). Now by Lemma (2.2), there exists an open set Δ in C^k containing $D \cup T_r$ and a function g which is an extension of f and is analytic in Δ. By Lemma (2.3), there exists a $\delta \in C^{k+}$ such that the conclusion of the Lemma (2.3) holds (with the present understanding of Δ). For any real $u, v > 0$ with $u < v$, let $A(u, v)$ denote that annulus $\{\zeta \in C : u < |\zeta| < v\}$. Now we define the function h on $P_{r+\delta}$ by

$$h(z) = (2\pi i)^{-k}\left(\prod_{j=1}^{k} \int_{|\xi_j| = r_j + \delta_j} d\xi_j\right) g(\xi) \prod_{j=1}^{k} (\xi_j - z_j)^{-1}, \tag{2.4}$$

for each $z \in P_{r+\delta}$. It follows from standard considerations (see § 2 of [5]) that h is analytic in $P_{r+\delta}$.

Now let us fix a $z \in P_{r+\delta}$. We observe, by Fubini's theorem concerning repeated integration (see e.g. [2]), that the right hand side of (2.4) remains the same under any change in order of integration. Let K be the set of all positive integers $\leqslant k$. Now for each $t \in K$, the integrand of the right hand side of (2.4) is an analytic function of ξ_t in $A(r_t - 2\delta_t, r_t + 2\delta_t)$, for fixed $\xi_j \in A(r_j - 2\delta_j, r_j + 2\delta_j)$, $\forall\, j \in K - \{t\}$.

From the above two observations, it follows that the right hand side of (2.4) remains the same even if we replace (successively) $r_j + \delta_j$ by $r_j - \delta_j$ for $j = 1, 2, \ldots, k$. Now, since $g = f$ on P_r, by Cauchy Integral formula (e.g. see [5], [9], [12]), it follows that $h(z) = f(z)$. Thus $h = f$ on $P_{r-\delta}$, and hence on P_r, because of the Identity Theorem. Now since h is analytic in $P_{r+\delta}$, by Taylor's Theorem (e.g. see [5], [9], [12]) its Taylor series about $\bar{O}$, which is the same as (2.1) converges in $P_{r+\delta}$. This contradicts the hypothesis that r is a boundary point of $|D|$ in C^{k+}. Hence the Theorem.

Lemma 2.4. Let $r \in C^{k+}$ and let δ be a positive real number. Then there exists a positive real δ' and positive integers M and N such that

$$P(\alpha, r/N) \subseteq P((1 - 1/M)\alpha, \delta' r) \subseteq P(\alpha, \delta r) \tag{2.5}$$

for each $\alpha \in T_r$.

Proof. We shall assume (which we may do without loss of generality) that $\delta < 1$. First we prove the lemma in the special case when $\alpha = r$. We choose a positive integer M such that $1 - \delta/2 < u = 1 - 1/M < 1$ and let us write $v = 1 - u = 1/M$. Now $v < \delta/2$ so that $v < \delta - v$. Let us choose a real δ' such that $v < \delta' < \delta - v$ and with this choice, it is easily seen that $\bar{P}(ur, \delta' r) \subseteq P(r, \delta r)$. Further, we observe that $r \in P(ur, \delta' r)$. This implies that there exists a positive integer N such that $\bar{P}(r, r/N) \subseteq P(ur, \delta' r)$. Hence follows the lemma when $\alpha = r$.

In the general case, let $\alpha = (r_1 \exp (i\theta_1), \ldots, r_k \exp (i\theta_k)) \in T_r$, where $\theta = (\theta_1, \ldots, \theta_k) \in \mathbb{R}^k$. We consider the mapping ϕ defined by

$$\phi(z) = (z_1 \exp (i\theta_1), \ldots, z_k \exp (i\theta_k)), \quad \forall\, z \in C^k$$

and observe that $\phi(P(tr, sr)) = P(t\alpha, sr)$ for all non-negative real t and all positive real s and deduce the lemma from its special case already established.

8.3. PRE-REQUISITES IN PROBABILITY

Throughout this section, we follow the standard notation, conventions and terminology of the Theory of Probability as detailed in [2], [4] and [6]. The usual Kolmogorov zero-one law for sequences (see e.g. [2], [4], [6]) is concerned with a sequence of independent random variables and for our purposes, we need the following generalisation of the same for nets (see [10] for definitions etc.).

Theory 3.1 (Generalised Kolmogorov zero-one Law-Lemma (1.3) of [7]). Let $X = \{X_n : n \in D\}$ be a net of real valued independent random variables, where $D = (E, \geqslant)$ is an infinite directed set with the property that to each $p \in E$, there exists a $q \in E$ such that $p \geqslant q$ but not $q \geqslant p$. Then every tail event of the net X has probability 0 or 1 and every tail function of the net X is a constant a.s. (almost surely).

8.4. GENERALISED STEINHAUS' THEOREM

In this section, we prove

Theorem 4.1. Let, as before,

$$\sum_{n\in I} c_n z^n (z \in C^k) \tag{4.1}$$

be a power series in C^k, where $\{c_n, n \in I\}$ is a net of (non-random) complex numbers. Let D be the (open) domain of convergence of (4.1), which is not all of C^k. Further, let $\{\phi_n, n \in I\}$ be a net of real-valued independent random variables on $(\Omega, \mathcal{F}, P)$ such that each ϕ_n is uniformly distributed in $[0, 2\pi)$. Then the random power series

$$\sum_{n\in I} c_n \exp (i\phi_n) z^n \quad (z \in C^k) \tag{4.2}$$

has D for its natural domain a.s., or more precisely the set A consisting of all $w \in \Omega$ such that the power series

$$\sum_{n \in I} c_n \exp (i\phi_n(w))z^n \quad (z \in C^k) \tag{4.3}$$

has D for its natural domain, is an event and it has probability 1.

It is convenient to divide a major part of the proof into some lemmas.

Lemma 4.1. Let D be as in Theorem (4.1) and let f be the sum function of (4.1) in D. Let α be a boundary point of D such that no component of α is 0 and let r denote $(|\alpha_1|, \ldots, |\alpha_k|)$. Let $u \in (0, 1)$ and $\delta \in (0, +\infty)$ and let E be the domain of convergence of the Taylor series of f about $u\alpha$. Then

$$\bar{P}(\alpha, \delta r) \subseteq E \Leftrightarrow (1 + \delta)\alpha \in E.$$

Proof. $\bar{P}(\alpha, \delta r) \subseteq E \Rightarrow (1 + \delta)\alpha \in \bar{P}(\alpha, \delta r) \subseteq E$ and this proves one part of the lemma.

To prove the other part, let us assume that $(1 + \delta)\alpha \in E$. Let $z \in \bar{P}(\alpha, \delta r)$. Now we have

$$\begin{aligned} |z_j - u\alpha_j| &\leqslant |z_j - \alpha_j| + |(1 - u)\alpha_j| \\ &\leqslant \delta r_j + (1 - u)r_j \\ &\leqslant |(1 + \delta)\alpha_j - u\alpha_j| \end{aligned}$$

for $j = 1, 2, \ldots, k$. Further E is a complete k-circular domain about $u\alpha$ (see [5] for definition). Hence it follows that $z \in E$. This completes the proof of the lemma.

Lemma 4.2. Let D and f be as in Lemma (4.1) and let $r \in C^{k+}$ be a boundary point of $|D|$. Let F be a nonempty compact subset of $\mathbf{R}^k$ and let S be a dense subset of F. Let χ be the map on $\mathbf{R}^k$ defined by

$$\chi(\theta) = (r_1 \exp (i\theta_1), \ldots, r_k \exp (i\theta_k)), \quad \forall\ \theta = (\theta_1, \ldots, \theta_k) \in \mathbf{R}^k.$$

Then every point of $\chi(F)$ is a regular point of the power series (4.1) iff there exist positive integers M and N such that

$$\limsup_{\|n\| \to +\infty} \left| \frac{1}{n!} \sum_{\substack{m \in I \\ m > n}} c_m \alpha^m \left(1 - \frac{1}{M}\right)^{\|m\|} \prod_{j=1}^{k} m_j(m_j - 1) \ldots (m_j - n_j + 1) \right|^{1/\|n\|} < \frac{(m-1)N}{M+N} \tag{4.3}$$

for each $\alpha \in \chi(S)$.

Proof of (4.3). For each $a \in D$, we write D_a to denote the domain of convergence of the Taylor series of f about a. It is easier to prove the second part of the lemma which we do first. Let us assume that there exist positive integers M and N such that (4.3) holds for each $\alpha \in \chi(S)$. Hence by Theorem (2.1), we have that $(1 + 1/N)\alpha \in D_{(1-1/M)\alpha}$ for each $\alpha \in \chi(S)$. Therefore, by Lemma (4.1), it follows that $\bar{P}(\alpha, r/N) \subseteq D_{(1-1/M)\alpha}$ for each $\alpha \in \chi(S)$, Now since S is dense in F and χ is continuous, it follows that $\chi(S)$ is dense in $\chi(F)$ and hence it follows that

$\chi(F) \subseteq \bigcup_{\alpha \in \chi(S)} P(\alpha, r/N)$ which shows that every point of $\chi(F)$ is a regular point of (4.1).

In order to prove the other part of the lemma, we assume that every point of $\chi(F)$ is a regular point of (4.1). Hence by Lemma (2.2), there exists an open set U of C^k containing $P_r \cup \chi(F)$ and a function g analytic in U and $g = f$ on P_r. Since F is compact, so is $\chi(F)$. Hence the distance d between $\chi(F)$ and $C^k - U$ is positive under the usual Euclidean norm of C^k(d being taken as $+\infty$ if $U = C^k$). Now we choose a real $\delta > 0$ such that $\delta\|r\| < d$. With this choice, we have that $\bar{P}(\alpha, \delta r) \subseteq U$, $\forall\ \alpha \in \chi(F)$. Therefore, by Lemma (2.4), there exist positive integers M and N and a real $\delta' \in (0, +\infty)$ such that

$$\bar{P}(\alpha, r/N) \subseteq \bar{P}((1 - 1/M)\alpha, \delta' r) \subseteq \bar{P}(\alpha, \delta r) \subseteq U$$

for each $\alpha \in \chi(F)$. But by Taylor's Theorem, the Taylor series of f about $(1 - 1/M)\alpha$ converges in $\bar{P}(\alpha, \delta r)$. Consequently we have that $\bar{P}(\alpha, r/N) \subseteq D_{(1-1/M)\alpha}$ for each $\alpha \in \chi(F)$. Now by Lemma (4.1), it follows that $(1 + 1/M)\alpha \in D_{(1-1/M)\alpha}$ for each $\alpha \in \chi(F)$. Hence by Theorem (2.1), it follows that (4.3) holds for each $\alpha \in \chi(F)$ and this completes the proof of the lemma.

Lemma 4.3. Let F and χ be as in Lemma (4.3) and let ϕ_n's be as in Theorem (4.1). Let R denote the set of all $\omega \in \Omega$ such that every point of $\chi(F)$ is a regular point of the series (4.3). Then R is an event and has probability either 0 or 1.

Proof. For the present and future use, let T denote the unit circle $\{z \in C : |z| = 1\}$ with the measure space structure induced by the transformation $\theta \to \exp(i\theta)$ for each $\theta \in [0, 2\pi)$, where the measure space under consideration consists of $[0, 2\pi)$, the family of all Borel subsets of $[0, 2\pi)$ and the Lebesgue measure $\div\ 2\pi$, restricted to the family. Let T^I denote the Cartesian product measure space $\underset{n \in I}{X}\ T_n$, where $T_n = T$, $\forall\ n \in I$. For each positive integer M, let H_M denote the extended real valued function on $T^I \times \mathbb{R}^k$, given by

$$H_M(x, \theta) =$$

$$\limsup_{\|n\| \to +\infty} \left| \frac{1}{n!} \sum_{\substack{m \in I \\ m > n}} c_m x_m r^m \left(1 - \frac{1}{M}\right)^{\|m\|} \prod_{j=1}^{k} m_j(m_j - 1) \ldots (m_j - n_j + 1) \right.$$
$$\left. \times \exp(i\theta_j m_j) \right|^{1/\|n\|}$$

for each $x = \{x_n;\ n \in I\} \in T^I$ and $\theta = (\theta_1, \ldots, \theta_k) \in \mathbb{R}^k$. (The involved series are all absolutely convergent, since all the derived series of (4.1) converge absolutely in D). Since each term of the series used to define H_M specifies a measurable function on $T \times \mathbb{R}^k$, the series defines a measurable function on $T^I \times \mathbb{R}^k$. Therefore, for each positive integer M, H_M is a measurable function on $T^I \times \mathbb{R}^k$, since it is the limit superior of a countable net of such measurable functions. Now let X denote the random object $\{\exp(i\phi_n); n \in I\}$ and let S be a countable dense subset of F. Hence by Lemma (4.2), it follows that

$$R = \bigcup_{M, N=1}^{+\infty} \bigcap_{\theta \in S} \left(H_M(X, \theta) < \frac{(M-1)N}{M+N} \right).$$

Now by the generalised Kolomogorov zero-one law (Theorem 3.1), each of the events involved in the intersection above has probability either 0 or 1 and so is the case with R. This completes the proof of the lemma.

Lemma 4.4. Let D and ϕ_n's be as in Theorem (4.1) and let $r \in C^{k+}$ be a boundary point of $|D|$. Further let T_r be the polycircle specified by

$$\{(r_1 \exp(i\theta_1), \ldots, r_k r^{i\theta_k}) : \theta_j \in [0, 2\pi) \quad \text{for } j = 1, 2, \ldots, k\}$$

Then the set A_r consisting of all $\omega \in \Omega$ such that every point of T_r is a singular point for (4.3) is an event and it has probability 1.

Proof. Let $N = N^k$ denote the set of all ordered k-tuples of positive integers and for each $q = (q_1, \ldots, q_k) \in N$, let N_q denote $\{t \in N : t \leqslant q\}$. For each $q \in N$ and $t \in N_q$, let $B_{q,t}$ denote the polyarc

$$\{(r_1 \exp(i\theta_1), \ldots, r_k \exp(i\theta_k)) : \theta_j \in E_{q,\, t_j} \quad \text{for } j = 1, 2, \ldots, k\}$$

where $E_{q,\, t_j} = [2\pi(t_j - 1)/q_j, 2\pi t_j/q_j]$ for $j = 1, 2, \ldots, k$ and further let $A_{q,t}$ denote the set of all $\omega \in \Omega$ such that every point of $B_{q,t}$ is a regular point for (4.3). Now for each $q \in N$ and $t \in N_q$, let us write $A'_{q,t}$ to denote $\Omega - A_{q,t}$ and $w_{q,t}$ for $P(A'_{q,t})$. Hence by Lemma (4.3) it follows that $A'_{q,t}$ is an event and $w_{q,t} = 0$ or 1 for each $q \in N$ and $t \in N_q$. Now for each $q \in N$,

$$\sum_{t \in N_q} w_{q,t} \geqslant P(\bigcup_{t \in N_q} A'_{q,t}).$$

Hence by Theorem (2.1), it follows that

$$\sum_{t \in N_q} w_{q,t} \geqslant 1, \quad \forall\, q \in N. \tag{4.5}$$

We now fix $q \in N$ and show that $w_{q,t}$ is independent of t ($t \in N_q$). For each $n \in I$, since ϕ_n is uniformly distributed in $[0, 2\pi)$, it follows that $\exp(i\phi_n)$ is uniformly distributed in the unit circle T with notation as in the proof of Lemma (4.3). Now, because of the independence of the ϕ_n's, for any real non-random net $u = \{u_n, n \in I\}$, the random objects $X = \{\exp(i\phi_n), n \in I\}$ and $X_u = \{\exp(i(\phi_n + u_n)), n \in I\}$ have the same distribution. Hence in the notation of Lemma (4.3), for each $\theta \in \mathbf{R}^k$ and each positive integer M, $H_M(X, \theta)$ and $H_M(X_u, \theta)$ have the same distribution. Now expressing $1 - w_{q,t}(t \in N_q)$ in terms of the $H_M(X, \theta)$'s as in the proof of Lemma (4.3), we observe that $w_{q,t} = w_{q,s}$ for all $t, s \in N_q$. This together with (4.5) implies that $w_{q,t} = 1$, $\forall\, t \in N_q$. Thus $P(A'_{q,t}) = 1$, $\forall\, q \in N$ and $t \in N_q$. Now we observe that $A = \bigcap_{\substack{q \in N \\ t \in N_q}} A'_{q,}t$ and hence A_r is an event with probability 1. This completes the proof of the lemma.

Proof of Theorem 4.1. Let U be a countable dense subset of the boundary of $|D|$ in C^{k+} and let A_r be as in Lemma (4.4) for each $r \in U$. Since U is countable and Lemma (4.4) implies that A_r is an event with $P(A_r) = 1$, $\forall\ r \in U$, in order to complete the proof of the theorem, it is sufficient to prove that $A = \bigcap_{r \in U} A_r$. It is easily seen that $A \subseteq \bigcap_{r \in U} A_r$. Now let $\omega \in \bigcap_{r \in U} A_r$. This implies that every point of $\bigcup_{r \in U} T_r$ (with T_r as in Lemma (4.4)) is a singular point of (4.3). Since the set of singular points of (4.3) contains $\bigcup_{r \in U} T_r$ as a dense subset, it follows that $\omega \in A$. Therefore, we have that $\bigcap_{r \in U} A_r \subseteq A$. This completes the proof of the theorem.

Remark 4.6. The extent to which the conditions on the net $\{\phi_n, n \in I\}$ in Theorem (4.1) could be relaxed is of interest. However there are some obvious (but possibly worth noticing) relaxations. For example, if μ is the product measure λ^I, where λ is the Lebesgue measure in $\mathbb{R}$ restricted to $[0, 2\pi)$, the Theorem (4.1) holds with the last sentence of its hypothesis weakend as "let $\{\phi_n, n \in I\}$ be a random object into $[0, 2\pi)^I$ whose distribution is absolutely continuous with respect to μ".

8.5. APPLICATION TO THE THEORY OF DOMAINS OF HOLOMORPHY

It is well-known that every domain in C is a domain of holomorphy (e.g. see [1], [12]), while this is false for domains in C^k, when $k \geqslant 2$. However, we have the well-known (e.g. see [12]).

Theorem 5.1. Let, as before, D be the domain of convergence of the power series (2.1), which is not all of C^k. Then D is a domain of holomorphy.

It is interesting that such a result admits a probabilistic proof which we shall now give.

Proof of (5.1). Let Λ be the measure space $([0, 2\pi), \mathcal{B}, \mu)$ where $\mathcal{B}$ is the family of all Borel subsets of $[0, 2\pi)$ and μ is the Lebsgue measure $\div\, 2\pi$ restricted to $\mathcal{B}$. We take the probability space $(\Omega, \mathcal{K}, P)$ as the Cartesian product measure space $\underset{n \in I}{X} \Lambda_n$, where $\Lambda_n = \Lambda$, $\forall n \in I$. For each $n \in I$, let ϕ_n denote the projection of Ω $(=[0, 2\pi)^I)$ onto the nth component space so that $\phi_n(\omega) = \omega_n$, $\forall \omega = \{\omega_t', t \in I\} \in \Omega$. It is now clear that ϕ_n's are independently and uniformly distributed in $[0, 2\pi)$. Therefore, by Theorem (4.1), there exists $\omega \in \Omega$ such that the series (4.3) has D for its natural domain. Hence D is a domain of holomorphy and this completes the proof of the theorem.

REFERENCES

1. L. V. Ahlfors (1966): *Complex Analysis*, McGraw-Hill Book Company.
2. R. B. Ash (1972): *Real Analysis and Probability*, Academic Press.

3. A. T. Bharucha-Reid and M. Sambandham: *Random polynomials*, Academic Press.

4. J. L. Doob (1965): *Stochastic Processes*, John Wiley.

5. B. A. Fuks (1963): Introduction to the Theory of Analytic Functions of Several Complex Variables, *Amer. Math. Soc.*

6. I. I. Gikhman and A. V. Skorokhod (1969): *Introduction to the Theory of Random Processes*, W. B. Saunders Company.

7. J. Gopala Krishna (1979): Power Series in One and Several Complex Variables with independent coefficients. Proceedings of All India Seminar on Hyperbolic Complex Analysis, *Publication of the Ramanujan Institute*, **No. 4**, pp. 180–188, University of Madras.

8. J. Gopala Krishna and I. H. Nagaraja Rao (1973): On orders and types of an entire function over C_k, *Jour. Aust. Math. Soc.*, **15**, pp. 393–408.

9. Lars Hormander (1973): *An Introduction to Complex Analysis in Several Variables*, North-Holland Publishing Company.

10. J. L. Kelley (1969): *General Topology*, Affiliated East West Press, Delhi.

11. Eugene Luckacs (1975): *Stochastic Convergence*—Second Edition, Academic Press.

12. Raghavan Narasimhan (1971): *Several Complex Variables*—Chicago Lectures in Mathematics—The Univ. of Chicago Press.

13. W. Rudin (1974): *Real and Complex Analysis*—Second Edition, Tata McGraw-Hill Publishing Company.

14. C. Ryll-Nardzewski (1953): D. Blackwell's Conjecture on Power Series with Random Coefficients, *Studia Mathematica*, **Tom XIII**, pp. 30–36.

15. H. Steinhaus (1930): *Über die Wahrscheinlickeit dofür, duβ der Konvergenz kreis einer Potenzreihe ihre natürliche Grenzeist*, Math. Zeitschrift, **Vol. 31**, pp. 408–416.

9

Spectral Class of Processes

C. S. K. Bhagavan
Andhra University, Waltair, A.P.

9.1 INTRODUCTION

An indexed set of random variables is known as a stochastic process. The indexing set is known as time and is taken to be infinite. The process is distinguished as discrete or continuous according as the index set is discrete or continuous. It is usual to take index set as $\{0, \pm 1, \pm 2, \ldots\}$ in the discrete case and as $(-\infty, \infty)$ in the continuous case. We shall treat the univariate discrete case in considerable detail for its ease and with occasional remarks about the continuous case.

Developments in stochastic processes took place in two main streams—one along the lines of probability structures and the other along the lines of moments, specifically the first two moments which are assumed to exist finitely. Here we shall follow the stream of moments and collect together the class of processes possessing a spectrum, a concept explained in the next section.

9.2 STATIONARY PROCESS

One basic assumption that led to considerable development generalising the old case is stationarity. It concerns the first two moments. It is defined as the second moment being a function of the time lag. Often the first moment is taken as time free. In this case, the process is called wide sense stationary (wss). This further restriction helps inference. Formally, wide sense stationarity means

$$m(t) = Fx(t) = m$$

$$C(t, u) = E[\overline{X(t) - m(t)}\{X(t+k) - m(t+k)\}]$$
$$= r(k),$$

bar denoting complex conjugate. $k = 0$ implies the constancy of the variance.

This stationarity was introduced by Khintchine in 1934. This is also known as weak stationarity. Against this there is a concept of strict stationarity which requires that the probability structures be invariant over time shifts, i.e. the probability distribution of $\{X(t_1 + k), X(t_2 + k), \ldots X(t_n + k)\}$ is the same as that of $\{X(t_1), X(t_2), \ldots X(t_n)\}$. The two concepts (weak and strict) coincide for a Gaussian process. The strict sense stationarity is used here rarely. Unless otherwise stated, stationarity in the following discussion means weak (wide sense) stationarity.

The discussion on wide sense stationary processes developed into two streams—one called domain analysis and the other called the frequency domain analysis. The frequency domain analysis is a speciality of the processes which takes place away from the stochastic action while time domain analysis is a sort of a carry forward of the earlier ideas in the stochastic arena. The frequency domain approach also serves as a unification of the discrete and continuous time domains by means of a "parameter" called frequency. That a frequency approach for stationary processes is possible is borne out by the well known spectral representation theorem by Bochner and later by Khintchine (1934) in the continuous parameter case and by Herglotz (1911) in the discrete parameter case. The representation theorem says that for a weak sense stationary process, the covariance function assumed continuous, can be represented as a Fourier transform

$$r(k) = \int_w e^{iks}\, dF(s), \tag{1}$$

where w lies in

$$[-\pi, \pi] \text{ or } (-\infty, \infty)$$

according as the time parameter is discrete or continuous and $F(\cdot)$ is a non-negative, non-decreasing and bounded function. The function $F(\cdot)$ is called spectrum of the process. This representation has similarity with that of the characteristic function. This spectrum is unique for a stationary process, while the converse situation may not be "true". This is because one can get back the second moments and these second moments may not be sufficient to give the process except in cases like normal. However, the representation can play the role similar to that of a characteristic function.

Looking at the Fourier representation of the second moments, one naturally raises the question as to whether a similar representation holds for the process variates as well. This has been answered through the pioneering Spectral Representation Theorem of Cramér in 1942, saying that for a stationary process, with continuous covariance,

$$X(t) - m(t) = \int_w e^{its}\, dZ(s), \tag{2}$$

where w is as explained before and Z is an additive set function with

$$Z(S_1 + S_2) = Z(S_1) + Z(S_2);\ S_1, S_2 \in \text{the Borel class of sets from } w;\ S_1S_2 = \phi,$$

$$EZ(S) = 0$$

$$E\overline{Z(S_1)}Z(S_2) = 0$$

This $Z(S)$ is a process of the continuous type, called the spectral process and is a process of orthogonal increments. Thus, this result has opened up a two way approach regarding the process and its second moments. In fact, it has provided a new tool of analysis of stationary processes.

The power of these spectral representations spread in another direction as well. The nature of these Fourier representations provides for the presence of discontinuities (jumps) in the spectrum and the spectral process. As has already been remarked the function $F(s)$ occurring in the covariance representation is to be viewed as distribution function. As is well known, this can be broken up into three components—one continuous part, another discrete part and the third called singular part with density zero. In practice the singular part is either combined with the continuous part or is assumed absent, without making specific mention about this. Thus only two components (the continuous part and discrete part) will be given prominence in the analysis. In fact, in the analysis of stationary processes the discrete part plays a dominant role. This is due to the following considerations:

Introducing the break-up of discrete and continuous parts, the covariance representation can be written as

$$r(k) = \int_W e^{iks} f(s)\, ds + \sum_r e^{iks} J_r$$

and correspondingly, the process representation in the form

$$x(t) - m(t) = \int_W e^{its}\, dz(s) + \sum_r e^{its} \tilde{z}_r$$

where $f(s)$ denotes the density of the associated continuous part, summations are over the possible discontinuities (which are at most enumerable) and J_r, $\tilde{z}_r$ are the associated jump parts. From this, one significant feature seen is that a discontinuity in the spectrum is induced by a periodic component in the process. *Indeed, this fact is useful as a detector of the presence of periodic components in the process.* This observation will be quite helpful in applications when one is analysing a realisation (sample observation) of the process, gathered over few time instants. These few realised values are known as time series. Time series are observed to display periodic nature. A device through which the periodic nature of time series is analysed is the periodogram analysis. Here we plot the function called the intensity function $I(s)$ or its modifications [see Anderson (1971)] against chosen values of s. The form of the intensity function is $I(s) = \{A^2(s) + B^2(s)\}^{1/2}$ with

$$A(s) = \frac{2}{n} \sum_{t=1}^{n} X(t) \cos \frac{2\pi t}{s};$$

$$B(s) = \frac{2}{n} \sum_{t=1}^{n} X(t) \sin \frac{2\pi t}{s}.$$

The nature of the periodogram is that it runs close to the X-axis except that there are sudden peaks at points corresponding to the jumps of the spectrum. Thus, *identifying the jumps in the spectrum, forms an aim of time series analysis.* Clearly this becomes possible through the spectral representations:

The prominent problems solved using this spectral representation theory are:

(a) Effects of smoothing.
(b) Decomposition into deterministic and nondeterministic components.
(c) Solving difference equations of the auto-regressive and/or moving average types, either taking a single equation or system of equations.
(d) Prediction and filtering.
(e) Estimation of the parameters like the mean value, auto-covariance.
(f) Estimation of the spectral function.
(g) Some limit theorems, called ergodic theorems concerning the limiting behaviour of the equi-weighted averages of the process variates taking the three basic modes of convergence (convergence in probability, in mean-square and almost sure).
(h) Maximum likelihood theory and Neyman-Pearson theory of testing—these in limited situations, and
(i) Generalisations to vector and multiparameter processes—these in a limited way.

The material on these various aspects of stationary processes can be got from the works of Anderson (1971), Avadhani (1969, 70), Bartlett (1966), Box and Jenkins (1970), Cramér and Leadbetter (1967), Grenander (1950), Grenander and Rosenblatt (1957), Loéve (1963), V. K. Murthy (1961 and subsequent papers), Medhi (1982), Nagabhushanam (1951, 1952), Nagabhushanam and Bhagavan (1962), Papoulis (1965), Parzan (1965), Venkataraman (1968 and the subsequent papers on the subject), Wiener (1942), Wold (1938) and Wold and Jureen (1964), among others.

9.3 THE SPECTRAL CLASS OF PROCESSES

The next step of advancement has naturally been by way of generalising these representations. This means we have to expand the class of stationary processes without losing contact with these spectral representations. This expanded class of processes has come to be known as "non-stationary processes" a term which has to be properly understood in its true perspective, namely, as *a class of processes not necessarily stationary.* When we move away from stationarity, the nice mathematical structures clearly disappear and one has to build them up again in either of the directions of the spectral representation of the covariance and the process representation. The primary object behind these being the concept of spectrum. One has now to think of the spectrum in the generalised form. We have thus to look at what we are expecting of the spectral function.

Broadly, the following could be the requirements of such a function [see Loynes (1968) and Nagabhushanam (1970)]:

> non-negativity and additivity like mass or energy, unique determinability from the auto-covariance function by Fourier-Stieltjes transformation, possibility of possessing a jump part, determinability of the spectral transfer function when the process variates undergo a simple linear transformation, estimability of the spectral density from a single realisation of the process, and reduction to the usual spectral function when the process is specialised to be a weakly stationary process.

Loynes listed all requirements for a spectral function of a process and concluded that when the process is not stationary, there does not seem to exist a spectral function satisfying all the requirements. Then what can be done is to seek a spectral function suitable for the particular enquiry, on hand. Thus the attempts in this area become diverse and have broadly developed into two streams—one taking a start from the process representation, another from the covariance.

The pioneering attempts of generalisation were by Karhunen (1942) and Loéve [see Loéve (1963)]. Their attempts were aimed at generalising the process representation (2). Karhunen considered a general function $f(t, s)$ in place of e^{its}, while Loéve considered omitting the requirement of orthogonality of the spectral process. Loéve considered spectral process with covariance function of bounded variation and has shown that the representations work both ways. That is, the covariance will have the representation,

$$r(t, u) = \int_{w} e^{-its+iuv}\, dh(s, v),$$

h being the covariance function of the spectral process is assumed to be a function of bounded variation and the process will have the representation,

$$X(t) - m(t) = \int_{w} e^{its}\, dZ(s).$$

This class of processes include the stationary processes when the mass of h is concentrated along the diagonal ($s = u$) and is called *harmonizable processes.* These processes were studied extensively by Cramér [see Cramér (1961) and the subsequent papers]. For the continuous parameter harmonizable processes, Cramér defined the spectrum as

$$\int_{-\infty}^{s} \int_{-\infty}^{\infty} |dh(s, v)|$$

Using this he solved the problems of prediction, filtering and the non-deterministic nature of the process.

The too general nature of Karhunen's representation streamlines the considerations in only one path, viz., from the point of view of the process representation. Processes arising from one particular specialisation of $f(t, s)$ which when treated as function of s, possesses Fourier transform with a particular kind of behaviour, called *oscillatory processes.* These were studied extensively by Priestley and his associates [see Priestley (1965) and the subsequent articles on this]. For these processes, Priestley defined the spectrum as $|f(t, s)|^2 g(s)$, where $g(s)$ is the function induced by the spectral process. The speciality of Priestley's spectrum is that it is slowly varying with time and is called an evolutionary spectrum. Using this, he solved the problem of estimation of the spectrum, prediction, filtering and a test for stationarity as well. One special case of evolutionary spectrum has also been considered by Granger and Hatanaka (1964).

Thus the attempts of Loéve-Cramér and Priestley take a start from the process representation.

Another line of attempt takes a start from the covariances and tries to build up the concept of spectrum from the covariances. One of the pioneering attempts in this direction is due to Parzen (1961). He considered a process of the continuous parameter and employing the technique of the continuity theorem (for characteristic functions), built up the spectral representation theorem for what may be called the averaged covariances as

$$\lim_{T \to \infty} \frac{1}{T} \int_0^{T-k} r(t, t+k)\, dt, \; k \geqslant 0,$$

assumed to exist finitely. He needed moments upto the fourth order. This method clearly does not apply to the discrete processes when the integral is replaced by a suitable sum. Parzen also considered the question of estimation of the spectral function.

In the discrete parameter case, Herbst (1964) (see his subsequent papers also), considered processes of the form

$$X(t) = \sum_{J=0}^{P} a_J C_{t-J}\, \epsilon_{t-J}, \; t = 0, \pm 1, \pm 2, \ldots,$$

where ϵ_t is a real Gaussian stationary process of iid random variables, a_J's being constants and C_t's being such that

$$C_t^2 < Q < \infty \text{ for each } t, \text{ and}$$

$$\lim_{N \to \infty} \frac{1}{N} \sum_{t=1}^{N} C_t^2 < \infty, \text{ say } C_o.$$

Herbst defined the spectrum for these processes as

$$C_o \sum_{J=0}^{P} |a_J e^{iJs}|^2.$$

Earlier attempts on introducing spectra for processes other than stationary processes lacked in some respect or the other. For example, the spectrum defined by Fano (1950) does not include stationary processes in an essential way and that of Page (1952) need not be necessarily non-negative.

An approach that applies to discrete as well as continuous processes, that directly generalised that of the stationary processes, was made by Nagabhushanam and/or Bhagavan (1968, 1974), who took a start from the covariances and their averaged covariances

$$\left(\lim_{n\to\infty} \frac{1}{n} \sum_{\substack{t,u=1\\u-t=k}}^{n} r(t, u) \text{ in the discrete case}\right)$$

They defined a set of functions and obtained the spectral representation (1) for the averaged covariances through a limiting process. For example, the sequence of these functions in the discrete case is

$$F_n(s) = \frac{1}{2\pi n}\left\{(\pi + s) \sum_{t=1}^{n} r(t, t) + \sum_{\substack{-k+1\\k\neq 0}}^{k-1} \sum_{\substack{t,u=1\\u-t=k}}^{n} \frac{e^{-iks} - e^{ik\pi}}{-ik} r(t, u)\right\}$$

(compare with that in Granander and Rosenblatt).

The generality of this method clearly enables one to expand the class of stationary processes by bringing into the fold, more classes of processes. This expanded class of processes can reasonably be termed as the *spectral class of processes, i.e. class of processes possessing a spectral representation.* Some of the well-known processes that readily come into this class are—Harmonizable processes [Nagabhushanam (1969), Bhagavan (1974)], Branching processes [Bhagavan (1974) and some classes of Markov processes [Nagabhushanam (1975, 1976)].

Another feature that comes up as we move away from stationarity is that we will have differing directional spectra, i.e., spectra obtained by the forward or the backward time movements. An example where the directional spectra differ is the following process

$$X(t) = \begin{cases} X_1 e^{it\pi/2} + X_2 & \text{for } t \geqslant 0 \\ X_1 e^{-it\pi/2} + X_2 & \text{for } t < 0 \end{cases}$$

where X_1, X_2 are random variables with zero means, variances each unity and covariance 1/2. The forward spectrum for this consists of two jumps of magnitude unity at 0 and $\pi/2$ whereas the backward spectrum consists of two jumps of magnitude unity at 0 and $-\pi/2$ [see Nagabhushanam (1970)].

This situation leads us to a new concept: when the directional spectra coincide, the common spectrum is called the *overall spectrum*. It may be generally remarked that this overall spectrum exists when there is process representation. Thus, for example, harmonizable processes and oscillatory processes possess overall spectrum [see Nagabhushanam (1969), Bhagavan (1974), Bhagavan and Somaraju (1979)]. For example, the overall spectrum of the discrete harmonizable process is

$$F(s) = h(s, s) - h(-\pi, -\pi) + J(s),$$

where

$$J(s) = \begin{cases} 0 & \text{for } s < \pi \\ 2J & \text{for } s = \pi \end{cases}$$

J standing for the real part of the discrete mass of h at $(-\pi, \pi)$.

Further classes of processes were brought under the fold of this generalised spectrum. They arise as generalisation of the Karhunan representation wherein the spectral process is not necessarily orthogonal and by specialising on the integrand function $f(t, s)$.

One class of processes called almost harmonizable processes arises by requiring that the function $f(t, s)$ be almost periodic. This class of processes also arises as solutions of some difference and differential equations [see M. M. Rao (1976)]. This class is related to the oscillatory process of Priestley and include the class of processes considered by Kampe-de-Feriet and Frenkiel, $x \cdot x$ processes arising as signal plus noise, signal having periodic covariance and with stationary noise [see M. M. Rao (1976)].

Another class of processes arises by requiring that the function $f(t, s)$ has Fourier transform [see Bhagavan and Somaraju (1979)]. The overall spectrum exists in both these general cases.

It may thus be seen that the attempts of workers such as Parzen, Nagabhushanam and/or Bhagavan, Kampe-de-Feriet, M. M. Rao and Bhagavan and Somaraju in arriving at a spectral function go along the line of establishing the existence of a spectrum, while the other attempts were different. The attempts of Herbst also were along the line of Nagabhushanam and Bhagavan. Attempts of Nagabhushanam and Bhagavan are a straightforward generalisation of stationary processes and are intended to serve as a common binding factor and give a time free spectral function. Incidentally, it can be thought of a link between the two approaches mentioned.

9.4. ERGODIC THEOREMS

One use to which this generalised spectrum can readily be put to is in respect of the ergodic theorems. The question examined here is that of the convergence of the equi-weighted means (and its variants). These means are $\frac{1}{n}\sum_{t=1}^{n} X_t$ in the case of discrete parameter and $\frac{1}{T}\int_0^T X(t)\,dt$ in the continuous parameter case. Their convergence to a constant is of particular interest. Two particular modes of convergence, namely, mean square and almost sure convergence, have received good attention, for they give consistent estimation of parameters.

Looking at the background of the stationary case, we have the well-known result,

$$\frac{1}{n}\sum_{t=1}^{n} [X(t) - m(t)]$$

that converges in the mean-square to a random variable whose mean is zero and variance equalling the possible jump of the spectrum at the origin. This result is known as the statistical ergodic theorem (see

Granander and Rosehblatt). Thus the lack of spectral jump at the origin forms a necessary and sufficient condition for the average considered, to converge to zero. The form in which this result could be readily extended to general processes by Nagabhushanam and Bhagavan (1969) is that the lack of spectral jump at the origin is a sufficient condition for the mean-square convergence of the averages to zero. However, a direct generalisation of the statistical ergodic theorem in respect of classes of processes having a process representation has been achieved. These classes of processes are harmonizable processes [see Nagabhushanam (1969) and Bhagavan (1974)] and their further generalisation mentioned in the previous section [see Bhagavan and Somaraju (1979)]. Incidentally, analogous results are valid for oscillatory processes also as they are a particular case of the class of these processes.

Coming to the question of almost sure convergence of the averages considered, the result for stationary processes involving the spectrum, says (see Loéve) that the differentiability of the spectrum in a neighbourhood of the origin implies almost sure convergence considered. This result has been extended by Bhagavan and Somaraju (1982) in respect of the general class of processes considered, by requiring that the spectrum be differentiable and the derivative be a function of bounded variation in a neighbourhood of the origin.

When these results are interpreted as giving consistency of estimators, the parameters they are estimating need some clarification. This matter is trivial for widesense stationary processes as the parameter there is the constant mean. But in respect of other processes there is need to introduce a corresponding parameter. It is called the *pooled mean* defined in the discrete time case as

$$\lim_{n\to\infty} \frac{1}{n} \sum_{t=1}^{n} m(t) \text{ or } \lim_{T\to\infty} \frac{1}{T}\int_0^T m(t)\, dt \text{ in the continuous time case.}$$

This pooled mean has been shown to exist in respect of processes having process representation. These processes include stationary processes and harmonizable processes (see Bhagavan). Thus when the pooled mean exists, the convergence of the sample mean is to the pooled mean.

9.5 CONCLUDING REMARKS

The concept of spectrum of stationary processes having been generalised in a straightforward manner, further efforts have to be in the direction of enlisting more classes of processes, thus enlarging the presently known classes possessing spectrum. The various aspects of estimation of spectrum of these processes is another area in which work can progress. However, the conventional methods of spectral estimation go through in a simple way. The several aspects of the spectral representation inducing a process representation is another area in which work has to progress.

REFERENCES

1. Anderson, T. W. (1971): *Stationary Time Series Analysis*, Wiley.
2. Avadhani, T. V. (1969): Sampling Errors of Covariances and Correlations for K-dimensional Stationary Stochastic Processes, *Sankhyā*, **Series B**, Vol. 31 pp. 13–24.
3. Avadhani, T. V. (Jointly with C. B. Bell and M. A. Woodroof) (1970): Some non-parametric tests in Stochastic Processes, First International Symposium on Non-parametric Techniques, Bloomington (USA) 1969 published by the Cambridge Univ. Press (1970) under the Title "*Non-parametric Techniques in Statistical Inference*", pp. 215–258.
4. Bartlett, M.S., (1966): *An introduction to stochastic processes with special reference to methods and applications*, Cambridge Univ. Press.
5. Bhagavan, C. S. K. (1974): *Non-Stationary Processes, Spectra and Some ergodic theorems*, A. U. Press, Waltair, India.
6. Bhagavan, C. S. K., and P. Somaraju (1979): "*Spectrum of k-h class of non-stationary processes*", presented at the symposium on probability and Statistical Inference, Poona, India.
7. ——— (1982): "*An Individual ergodic theorem for a class of non-stationary processes*" Presented at the Symposium on Probability and Statistical Inference, Poona, India.
8. Box, G. E. P. and Jenkins, G. M. (1970): *Time Series analysis, Forecasting and Control*, Holden-Day.
9. Cramer, H. (1942): On Harmonic analysis in certain functional spaces, *Arkiv fur. Math. Astr. Fys.*, **28 B.**
10. ——— (1961) : On Some classes of non-stationary stochastic processes Proc. of the Fourth Berk. Sympo., **Vol. II**, Univ. of Calif. Press, 57–78.
11. Cramer, H. and Leadbetter, M. R. (1967): *Stationary and related stochastic processes*, Wiley.
12. Fano, R. M. (1950): Short time auto-correlation functions and power spectra—*J. Acoust. Soc. Amer.*, **22**, 546–550.
13. Granander, U. (1950): Stochastic Processes and Statistical Inference, *Arkiv fur Mathematic*, 195–277.
14. Granander, U. and Rosenblatt, M. (1957): *Statistical analysis of Stationary Time series*, Wiley.
15. Granger, C. W. J. and Hatanaka, M. (1964): *Spectral analysis of economic time series*, Princeton.
16. Herbst, L. J. (1964): Spectral analysis in the presence of variance fluctuations—*J. Roy. Stat. Soc.*, **Series B, 21**, 354–360.
17. Herglotz, G. (1911): *Uber Potenzreihen mit positiven reellem Teil im Einheitskreis Ber. Verk Kgl.* **63**, 501–511.
18. Karhunen, K. (1947): *Uber ltneare Methods in der Wahrschein-lichkeitscechnug—Ann. Ac. Sci. Fennicac*, **AI 37**, Helsinki.
19. Khintchine, A. I. (1934): *Korrelationstheorie der stationaren stochastiche processe—Math. Annalen*, **109**, 604–615.
20. Loeve, M. (1963): Probability theory—D. Van Nostrand Co., Third edn.

21. Loynes, R. M. (1968): On the concept of the spectrum for non-stationary processes—*J. Roy. Stat. Soc.*, **Series B, 30, I–30.**

22. Medhi, J. (1982): ***Stochastic Processes***, Wiley Eastern.

23. Murty V. K. (1961): On the estimation of the spectral density, ***Annals of Mathematical Statistics*, 32.**

24. Nagabhushanam, K. (1951): The Primary process of a smoothing relation, ***Arkivfur** Mathematik*, 421–488.

25. ——— (1952): Some aspects of stationary time series, ***Sankhyā*, 12**, 109–116.

26. Nagabhushanam, K. (1969): Harmonizable processes and mean ergodic theorem, *Sankhyā*, **A31**, 413–420.

27. ——— (1970): Spectral concepts in non-stationary processes-unpublished proceedings of the second Mastech conference on Probability and Statistics, Madras, India.

28. ——— (1975): Mean ergodic theorem for an irreducible time homogeneous Markov chain with a finite number of states, *Current Science*.

29. ——— (1976): Some ergodic theorems for non-stationary Markov processes—Lecture to the symposium on advanced probability and statistics at Annamalainagar, India.

30. Nagabhushanam, K. and Bhagavan, CSK (1962): Decomosition of a class of stationary processes, *Current Science*, 1962.

31. ——— (1968): Non-Stationary processes and spectrum *Canadian Journal of Mathematics* **20**, 1203–1206.

32. ——— (1969): A mean ergodic theorem for a class of non-stationary processes, *Sankhyā*, **A 31**, 421–426.

33. Papoulis, A. (1965): ***Probability, random variables and stochastic processes***, McGraw Hill.

34. Page, C. H. (1952): Instantaneous power spectra—***J. Apply. Phy.*, 23**, 103–106.

35. Parzen, E. (1961): See Time series analysis papers—Holdenday, **1967**, papers **10** and **11**.

36. ——— (1965): ***Modern Probability theory and its applications***, McGraw-Hill.

37. Rao, M. M. (1976): Covariance analysis of some non-stationary time series, Technical report, *AFFDL TR* **76**, 126 Aug. 1974–1976.

38. Venkataraman, K. N. (1968): Some Limit theories on a linear explosive stochastic difference equations with a constant term and their statistical applications, *Sankhyā* **A 30, 51–74.**

39. Wiener, N. (1949): ***The extrapolation, interpolation and smoothing of stationary processes***, Wiley.

40. Wold, H. (1938): ***A Study in the Analysis of stationary time series***, Uppsala.

41. Wold, H. and Jureen (1964): ***Demand analysis***, Wiley.

10

Selecting a Subset Containing the 'Best' Negative Binomial Population*

N. S. Bartlett and Z. Govindarajulu

The Standard Oil Company
and
University of Kentucky

10.1 INTRODUCTION

In many practical situations, one wishes to decide which of a number of competing processes produces the largest proportion of good units. For instance, a buyer of ball-bearings specifies tolerance limits on the diameter of the bearings, and will buy from those manufacturers whose processes produce the highest proportion of items whose diameters are within those tolerance limits. Often one destroys the item being tested, as when one is concerned with fuses or the breaking strength of materials. In such a case one is clearly interested in a sampling plan for which the expected number of destroyed (bad) items is small. One plan which is known to have a small average number of defectives and small average sample size, particularly when the proportion p of good items produced by the process is fairly large, is called inverse binomial sampling. Under this plan one continues sampling, that is, continues testing items, until some predetermined number r of successful events has occurred.

In some situations the experimenter will wish to choose just one population as best. The tester in a commercial situation is likely to be willing, if not anxious, to choose more than one supplier for a given product, so that his own plant is not forced to shut down because of a strike or other major catastrophe affecting his only supplier, or so that he fosters continual price competition among his suppliers. Consequently, both the indifference zone and subset approach to the selection problem are of practical interest. However, in this paper we will confine ourselves to the subset approach.

---

*This research was supported by the Standard Oil Company (Ohio), the National Science Foundation Grants NSF-GP 569 and NSF-GP 5664, and Air Force Office of Scientific Research under Research Grant AF-AFOSR 741-65.

Bartlett (1965), in his unpublished dissertation considered this problem. An alternate subset approach has been considered by Gupta and Negal (1971, pp. 91–92).

10.2 NOTATION AND FORMAL STATEMENT OF THE SUBSET SELECTION PROCEDURE

Let Y denote the number of failures preceding the first success, when the probability of success in each of a series of independent trials is p. The random variable X has the geometric distribution and the random variable $X = \sum_{i=1}^{r} Y_i$ has the negative binomial distribution with parameters r and p, with

$$P(X = x; r, p) = \begin{bmatrix} x + r - 1 \\ x \end{bmatrix} p^r q^x, \quad x = 0, 1, 2, \ldots, q = 1 - p,$$

where Y_i denotes the number of failures between $(i - 1)$st and ith success $(i = 1, 2, \ldots, r)$. The random variable X represents the number of failures preceding the rth success when the probability of success on each independent trial is p. We may, therefore, think of the selection problem as being one of picking a subset containing the geometric population with the largest p value, where we take a sample of size r_i from population π_i, $i = 1, \ldots, k$, or alternatively and equivalently as one of picking a subset containing the negative binomial population (with any parameter r) having the largest p value, based on a single observation from each of the k populations.

Suppose then that X_i, $i = 1, \ldots, k$ are independent random samples from populations π_i, which have negative binomial distributions with parameters r_i, p_i, and that one wishes to select a subset of the populations which is to contain the population with the largest value of p with probability at least P^*, regardless of the true values of the p_i's. Any population π_i whose parameter p_i is equal to the largest of the values $p_1, \ldots, p_k$ will be called a best population. A correct selection (CS) will be said to have occurred if the subset selected as the result of an experiment contains at least one best population. Let $\tilde{r} = \{r_1, \ldots, r_k\}$ and $\tilde{p} = \{p_1, \ldots, p_k\}$. The probability of a correct selection using a procedure R, when k and $\tilde{r}$ are given, and when the true configuration of the population parameters is $\tilde{p}$, will be denoted by $P\{\text{CS}; k, \tilde{r}, \tilde{p}, R\}$. We define $P\{\text{CS}\}$, when $\tilde{p}$ satisfies

$$p_1 = \ldots = p_k \; (=p, \text{ say}) \tag{2.1}$$

to be the limiting value of $P\{\text{CS}\}$ as (2.1) is approached through parameter points for which one parameter value is strictly greater than all the others, which amounts to saying that when (2.1) holds, we consider one 'tagged' population as best.

Let $k^{-1} \leqslant P^* \leqslant 1$ be given. (Note that we can guarantee a probability

$P^* = 1/k$ of correct selection by picking one population at random, without experimentation, so that a value of P^* which is less than k^{-1} is of no practical interest.) After having defined a procedure $R(k, \tilde{r}, P^*)$ for all $k, \tilde{r}$ and P^* for which

$$P\{\text{CS}; k, \tilde{r}, \tilde{p}, R\} \geqslant P^*, \tag{2.2}$$

for all possible configuration $\tilde{p}$, one joint confidence statement that can be made, at confidence level P^*, is that the parameter values of the eliminated populations are all less than or equal to the smallest of the parameter values of the retained populations.

A criterion by which one can judge the relative merits of procedures which satisfy the basic requirement on $P\{\text{CS}\}$ is the expected size of the subset, $E(S; k, \tilde{r}, \tilde{p}, P^*, R(k, \tilde{r}, P^*))$. If one thinks of the problem in terms of the geometric distribution, in which case r_i is the sample size for population π_i, and if all the sample sizes are equal, then one can determine the common sample size r which is necessary in order to guarantee that, for a given parameter point $\tilde{p}_0$, and a given ϵ, $0 < \epsilon < k - 1$,

$$E(S; k, r, \tilde{p}_0 \;\; P^*, R(r)) \leqslant 1 + \epsilon. \tag{2.3}$$

Alternatively, one can determine r from the requirement that the infimum of $E(S)$ over some subset of the parameter space be less than or equal to $1 + \epsilon$.

Let the population with the jth smallest p value be assigned the rank j, $(j = 1, \ldots, k)$ and assume, without loss of generality and in order to avoid a more complicated notation, that $p_1 \leqslant p_2 \leqslant \ldots \leqslant p_k$. The ranks of the populations retained in the subset are random variables, and one may be interested in the sum of the ranks of the retained populations. One would tend to favour procedures for which the expected sum of ranks of the retained populations is relatively high, and for which the expected number of populations retained, $E(S)$, is relatively small. Let

$$P_i(k, \tilde{r}, \tilde{p}, R) = P \text{ (include } \pi_i \text{ in the subset)} \tag{2.4}$$

Then, the expected size of the subset is

$$E(S) = \sum_{i=1}^{k} P_i, \tag{2.5}$$

and the expected sum of ranks of the retained populations is

$$E(\text{sum of ranks}) = \sum_{i=1}^{k} iP_1.$$

10.3. THE SUBSET PROCEDURE R

The maximum likelihood estimate of p_i, when the random variable X_i has a negative binomial distribution with parameters r_i, p_i, is $p_i = r_i/(r_i + x_i)$, when x_i is the observed value of X_i. Hence, in the absence of any prior information about the p_i, $i = 1, \ldots, k$, one would tend to use a procedure which accepts or rejects populations according to the relative magnitudes of $p_1, \ldots, p_k$. Accordingly the procedure R is "retain population π_i in the selected subset if and only if

$$r_i/(r_i + x_i) \geqslant dr_j/(r_j + x_j), \quad \text{for all } j, \tag{3.1}$$

where $d = d(\tilde{r}, k, P^*)$ is a constant, $0 \leqslant d \leqslant 1$." Letting $b = d^{-1}$, the rule in (3.1) becomes "retain π_i in the selected subset if and only if

$$x_i \leqslant br_i(r_j)^{-1}x_j + (b - 1)r_i, \quad \text{for all } j, \tag{3.2}$$

where $b = b(k, \tilde{r}, P^*)$ is a constant $(1 \leqslant b < \infty)$."

The constant b is the smallest number for which the basic probability requirement (2.2) can be satisfied for all parameter points $\tilde{p}=\{p_1, \ldots, p_k\}$. Although we can, in principle, find b for any vector $\tilde{r} = \{r_1, \ldots, r_k\}$, we henceforth restrict our attention to the case $r_1 = \ldots = r_k$ $(=r$, say), in which case the inequality (3.2) becomes

$$x_i \leqslant bx_j + (b - 1)r, \quad \text{for all } j \ (1 \leqslant b < \infty). \tag{3.3}$$

Table 10.1 contains the values of $b(k, r, P^*)$ for selected values of k, r and P^*. To illustrate the use of the table, let us consider the following example.

Example 3.1. Let $k = 4$, $r = 2$ and $P^* = 0.95$. Let the observed values for X_1, X_2, X_3 and X_4 be 24, 5, 32, 0 respectively. From Table 10.1, we find that $b(4, 2, .95) = 11.76$ and consequently $bx_{\min} + (b - 1)r = (11.76)0 + (10.76)2 = 21.52$ and the experimenter retains the two populations which gave rise to the values 0 and 5 and can say with 95% confidence that one of the two retained populations has the largest p-value.

It is interesting to note that if one waits for the same number of successes, r, to occur in each population, having no reason a priori to favour one population over the others, and if one considers the p_i's to have, a priori, independent and identical beta distributions, then, following a Bayesian approach in the spirit of Guttman and Tiao, one is again led to base the rule upon the maximum likelihood estimator $r/(r + x)$. For, let the utility of the ith population be $U_i = U(\pi_i) = p_i$, so that in trying to maximize utility, one is looking for the population with the largest p-value. The assumption that the p_i's have a priori independent beta distributions, together with the assumption of independent samples from the k populations, assures that $U_1, \ldots, U_k$ are locally independent a posteriori, thus they may be analyzed individually. The likelihood function for p is

$$L(p; r, x) = \binom{x + r - 1}{x} p^r q^x = f(x; p, r).$$

Let the a priori beta distribution of p have density function

$$g(p) = \frac{(n_1 + n_2 - 1)!}{(n_1 - 1)!\,(n_2 - 1)!} p^{n_1-1} q^{n_2-1}, \; 0 \leqslant p \leqslant 1, \; q = 1 - p.$$

Table 10.1. *b*-Values needed to guarantee a probability of correct selection of at least P^* for any true configuration of parameters when procedure R is used with a common number r of observations from each of k geometric populations. The starred column gives b'-values for $r = 10$ using the normal approximation of Section 7.

$P^* = .75$

k \ r	1	2	3	4	5	6	7	8	9	10	*
2	3.00	2.07	1.79	1.64	1.56	1.50	1.45	1.41	1.39	1.36	1.35
3	6.00	3.16	2.48	2.17	1.98	1.86	1.77	1.70	1.65	1.61	1.57
4	9.00	4.02	2.97	2.52	2.26	2.09	1.97	1.08	1.81	1.75	1.70
5	12.00	4.74	3.37	2.79	2.47	2.25	2.11	2.00	1.92	1.85	1.79

$P^+ = .90$

k \ r	1	2	3	4	5	6	7	8	9	10
2	9.00	4.11	3.06	2.59	2.33	2.15	2.03	1.93	1.85	1.80
3	18.00	6.14	4.14	3.34	2.90	2.62	2.43	2.28	2.17	2.08
4	27.00	7.71	4.91	3.84	3.27	2.92	2.68	2.50	2.36	2.25
5	36.00	9.05	5.52	4.22	3.55	3.12	2.84	2.64	2.49	2.37

$P^* = .95$

k \ r	1	2	3	4	5	6	7	8	9	10	*
2	19.00	6.39	4.29	3.44	2.98	2.69	2.49	2.34	2.22	2.13	2.09
3	36.00	9.42	5.72	4.37	3.66	3.23	2.94	2.73	2.57	2.44	2.36
4	57.00	11.76	6.73	4.99	4.11	3.58	3.23	2.97	2.78	2.63	2.52
5	76.00	13.73	7.54	5.47	4.45	3.81	3.41	3.13	2.92	2.75	2.63

$P^* = .99$

k \ r	1	2	3	4	5	6	7	8	9	10
2	99.00	15.98	8.47	6.03	4.85	4.16	3.70	3.38	3.13	2.94
3	198.00	23.07	11.04	7.47	5.83	4.89	4.29	3.87	3.55	3.31
4	297.00	28.51	12.84	8.44	6.47	5.36	4.66	4.17	3.81	3.54
5	396.00	33.11	14.27	9.19	6.95	5.72	4.94	4.40	4.01	3.71

The posterior distribution of p, given x, is

$$g(p; x) = g(p)f(x; p)\left[\int_0^1 f(x; p)g(p)\,dp\right]^{-1}$$
$$= \frac{(r + x + n_1 + n_2 - 1)!}{(r + n_1 - 1)!\,(r + n_2 - 1)!}\,p^{r+n_1-1}q^{x+n_2-1},$$

which is again a beta distribution, and the posterior means of utility function U, given x, is

$$E(U; x) = \int_0^1 U(p)g(p; x)\,dp$$
$$= \int_0^1 pg(p; x)\,dp$$
$$= (r + n_1)/(r + x + n_1 + n_2).$$

Thus, one would retain those populations with small values of x, that is, large values of $r/(r + x)$.

10.4 THE PROBABILITY OF CORRECT SELECTION AND ITS INFIMUM

Recalling our assumption that $p_1 \leqslant p_2 \leqslant \ldots \leqslant p_k$ for simplicity of notation, and employing (3.3) we note that

$$P\{CS; k, r, b, \tilde{p}\} = P(X_k \leqslant bX_j + (b-1)r, \text{ for } j = 1, \ldots, k-1)$$
$$= \sum_{x=0}^{\infty}\binom{x+r-1}{x}p_k^r q_k^x \prod_{j=1}^{k-1}\sum_{y=f(x)}^{\infty}\binom{y+r-1}{y}p_j^r q_j^y, \quad (4.1)$$

where $f(x) = 0$ if $x < (b-1)r$, $f(x) = b^{-1}\{x-(b-1)r\}$ if this quantity is an integer, and $f(x) = [b^{-1}\{x-(b-1)r\}] + 1$ otherwise, where $[a]$ is the greatest integer less than or equal to a. Making use of the well-known identity relating the negative binomial cdf to the incomplete beta function, we can write (4.1) as

$$P\{CS; k, r, b, \tilde{p}\}$$
$$= \sum_{x=0}^{\infty}\binom{x+r-1}{x}p_k^r q_k^x \prod_{j=1}^{k-1}\left\{\frac{(f(x)+r-1)!}{(f(x)-1)!\,(r-1)!}\int_{p_j}^{1} y^{r-1}(1-y)^{f(x)-1}dy\right\}.$$

It is immediately evident from this representation that $P\{CS\}$ is a non-increasing function of p_j, for $j = 1, \ldots, k-1$, and consequently that $P\{CS\}$ is partially minimized as we let each p_j approach p_k from below, that is, a partial minimum of $P(CS\}$ occurs when (2.1) holds, and then

$$P\{CS; k, r, b, p\} = \sum_{x=0}^{\infty}\binom{x+r-1}{x}p^r q^x\left[\sum_{y=f(x)}^{\infty}\binom{y+r-1}{y}p^r q^y\right]^{k-1}$$
$$= \sum_{x=0}^{[(b-1)r]}\binom{x+r-1}{x}p^r q^x + \sum_{x=[(b-1)r]+1}^{\infty}\binom{x+r-1}{x}p^r q^x$$
$$\times\left[\sum_{y=f(x)}^{\infty}\binom{y+r-1}{y}p^r q^y\right]^{k-1} \quad (4.2)$$

Though an analytic proof seems to be very difficult to obtain, extensive computations convince one that the right-hand side of (4.2) is an increasing function of p for $0 < p < 1$. Thus $P\{\text{CS}\}$ is completely minimized when $p_1 = p_2 = \ldots = p_k = 0$, and b is the solution of

$$P\{\text{CS}; k, r, b, 0\} = P^* \tag{4.3}$$

Because every term in the infinite series expression (4.2) approaches zero as p approaches zero, we must derive a finite sum expression for $P\{\text{CS}\}$ in order to evaluate $P\{\text{CS}\}$ at $p = 0$. We will obtain in Theorem 4.1 such an expression under the assumption that b is an integer, and show that using this expression when b is not an integer leads to estimates of $b(k, r, P^*)$ which are at least approximately correct.

Theorem 4.1. Let b be a positive integer. Then
$P\{\text{CS}; k, r, b, p\} = F((b-1)r; r, p)$

$$+ \sum_{y_1=0}^{r-1} \cdots \sum_{y_{k-1}=0}^{r-1} \sum_{s=0}^{r-1+\Sigma y_i} \left\{ \frac{g(s, b, r, \tilde{y}) p^{r+\Sigma y_i} q^{(b+k-2)r+1-\Sigma y_i} P_s(q^{b+k-1})}{(r-1)! \prod_{i=1}^{k-1} y_i! \, [1 - q^{b+k-1}]^{s+1}} \right\} \tag{4.4}$$

where $F(x; r, p)$ is the cdf of a negative binomial distribution, $g(s, b, r, \tilde{y})$ is a positive integer, $\tilde{y} = \{y_1, \ldots, y_{k-1}\}$, and $P_s(x)$ is a polynomial of degree s in x (See Lemma in Appendix to the chapter). Furthermore,

$$\lim_{p \to 0} P\{\text{CS}; k, r, b, p\} = F(r-1, \ldots, r-1; r, bt, t, \ldots, t) \tag{4.5}$$

$$\begin{aligned}
&\lim_{p \to 0} \frac{d}{dp} P\{\text{CS}; k, r, b, p\} \\
&= \tfrac{1}{2}(k-1)t[(b+1-r)F(r-1, \ldots, r-1, ; r-1, bt, t, \ldots, t) \\
&\quad + (2r-b-1)F(r-2, r-1, r-1, \ldots, r-1; r, bt, t, \ldots t) \\
&\quad - rb^{-1}(b+k-2)F(r-3, r-1, r-1 \ldots, r-1; r+1, bt, t, \ldots t) \\
&\quad + rb^{-1}(k-2)F(r-2, r-2, r-1, \ldots, r-1; r+1, bt, t, \ldots, t)]
\end{aligned} \tag{4.6}$$

where $t = (b+k-1)^{-1}$ and where $F(x_1, \ldots, x_{k-1}; r, p, p_1, \ldots, p_{k-1})$ is the cdf of a negative multinomial distribution, with $p + \sum_{i=1}^{k-1} p_i = 1$, and we define $F = 0$, if any $x_i < 0$.

Proof. See Appendix to this chapter.

In the examples below we will display a non-integral value of b for which (4.5) holds. Let us consider two special cases.

Example 4.1(*i*): Let $r = 1$, so that we curtail sampling on any population as soon as the first success is observed. Then, from (A3), which is a special case of (4.5) we see that $b(k, 1, P^*)$ is approximately given by the solution of $b(b+k-1)^{-1} = P^*$, which yields $b = (k-1)P^*/(1-P^*)$.

If $P^*/(1 - P^*)$ is an integer, (as when $P^* = .75, .90, .95$, or $.99$) b is an integer. If $P^* = .60$, then $b = 1.5(k - 1)$. Let $k = 2$. Then

$$P\{\text{CS}; 2, 1, 1.5, p\} = \sum_{x=0}^{\infty} pq^{3x} \sum_{y=2x}^{\infty} pq^{y} + \sum_{x=0}^{\infty} \sum_{j=1}^{2} pq^{3x+j} \sum_{y=2x+1}^{\infty} pq^{y},$$

which, after straight-forward calculations, becomes

$$P\{\text{CS}; 2, 1, 1.5, p\} = p(1 + q^2 + q^3)/(1 - q^5),$$

so that

$$\lim_{p\to 0} P\{\text{CS}; 2, 1, 1.5, p\} = 3/5 = b/(b + 1)$$

so that (4.5) is exact in this case.

(ii) Let $k = 2$. Then (A6) becomes

$$\sum_{x=0}^{r-1} \binom{x + r - 1}{x} \left(\frac{b}{b + 1}\right)^r \left(\frac{1}{b + 1}\right)^x = P^*, \tag{4.7}$$

which can be expressed in terms of the incomplete beta function ratio as

$$I_{b/(b+1)}(r, r) = P^*, \text{ where } I_x(a, b) = \frac{(a + b - 1)!}{(a - 1)!\,(b - 1)!} \int_0^x u^{a-1}(1 - u)^{b-1} du.$$

The solution $x(r, P^*)$ of $I_x(r, r) = P^*$ may be found in Table 1 of Govindarajulu and Hubacker (1964), accurate to eight decimal places, for $r = 1(1)15$, $18(5)28$ and $P^* = .01, .025, .10, .25, .50, .75, .90, .975$ and $.99$. It also may be obtained, for more values of r and P^*, but with less accuracy, from Pearson's (1934) tables. Writing $x^* = x(r, P^*)$, we have $b(2, r, P^*) = x^*/(1 - x^*)$. We may also express (4.7) in terms of the cdf of the binomial distribution as

$$\sum_{x=0}^{r-1} \binom{2r - 1}{x} \left(\frac{b}{b + 1}\right)^x \left(\frac{1}{b + 1}\right)^{2r-1-x} = 1 - P^*,$$

and solve for $x^* = b/(b + 1)$ using tables of the binomial cdf, a list of which may be found in Pearson and Hartley (1956).

Note that if we define $b(p)$ as the smallest number for which $P\{\text{CS}; p, b\} \geqslant P^*$, then $b(p)$ can be regarded as the P^*-percentile of the statistic $V = \max\left[\frac{X_0 + r}{X_1 + r}, \ldots, \frac{X_0 + r}{X_{k-1} + r}\right]$, where $X_0, X_1, \ldots, X_{k-1}$ are independent random variables with negative binomial distributions with common parameters p and r, and the cdf of V, that is $P(V \leqslant b)$, is given by (4.2).

10.5 EXPECTED SIZE OF THE SUBSET

Let $r_1 = r_2 = \ldots = r_k = r$. Then with P_i defined by (2.4), we have

$$P_i(k, r, b, \tilde{p}) = P(X_i \leqslant bX_j + (b - 1)r, \text{ for all } j \neq 1)$$

$$= \sum_{x=0}^{\infty} \binom{x + r - 1}{x} p_i^r q_i^x \prod_{j\neq i} \left\{ \sum_{y=f(x)}^{\infty} \binom{y + r - 1}{y} p_j^r q_j^y \right\}, \tag{5.1}$$

where $f(x)$ is defined in (4.1). When $i = k$, (5.1) and (4.1) are equivalent. By virtue of (2.5) and (5.1), we have

$$E(S; k, r, b\ (k, r, P^*), \tilde{p})$$

$$= \sum_{i=1}^{k} \sum_{x=0}^{\infty} \binom{x+r-1}{x} p_i^r q_i^x \prod_{j \neq i} \left\{ \sum_{y=f(x)}^{\infty} \binom{y+r-1}{y} p_j^r q_j^y \right\}, \quad (5.2)$$

One set of parameter configurations of particular interest is that in which

$$p_1 = p_2 = \ldots = p_{k-1} = p,\ p_k = p + \delta, \quad (5.3)$$

for a given $\delta > 0$. We see from (5.2) that in this case

$$E(S; k, r, b, p, \delta)$$

$$= \sum_{x=0}^{\infty} \binom{x+r-1}{x} (p+\delta)^r (q-\delta)^x \left[\sum_{y=f(x)}^{\infty} \binom{y+r-1}{y} p^r q^y \right]^{k-1}$$

$$+ (k-1) \sum_{x=0}^{\infty} \binom{x+r-1}{x} p^r q^x \left[\sum_{y=f(x)}^{\infty} \binom{y+r-1}{y} (p+\delta)^r (q-\delta)^y \right]$$

$$\times \left[\sum_{y=f(x)}^{\infty} \binom{y+r-1}{y} p^r q^y \right]^{k-2}. \quad (5.4)$$

The first term on the right-hand side of (5.4) is the probability of correct selection for the parameter configuration (5.3). When $p = 0$, we have

$$E(S; k, r, b, 0, \delta) = \sum_{x=0}^{\infty} \binom{x+r-1}{x} \delta^r (1-\delta)^x = 1,$$

that is, we are certain to make the correct selection, while if $p = 1 - \delta$,

$$E(S; k, r, b, 1-\delta, \delta) = 1 + (k-1) \sum_{x=0}^{[(b-1)r]} \binom{x+r-1}{x} (1-\delta)^r \delta^x. \quad (5.5)$$

If $\delta = 0$, then for general p

$$E(S; k, r, b, p, 0) = kP\{\text{CS}; k, r, b(k, r, P^*), p\} \geq kP^*,\ 0 \leq p \leq 1. \quad (5.6)$$

From either (5.5) or (5.6) it is evident that $E(S; k, r, b, 1, 0) = k$, so that k is the least upper bound on values of $E(S)$.

Values of $E(S)$ for selected values of k, r, p, δ and P^* are provided in Table 10.2. These values were computed by truncating the summation in x in (5.4) to $(0, M)$.

Table 10.2 Expected Proportion of Populations Retained in the Selected Subset by Procedure R

For the true configuration $p_1 = p_2 = \ldots = p_{k-1} = p$; $p_k = p + \delta$ for selected values of k, r, P^*, δ and $p + \delta$

	$p^* = .75$									
	$k = 2$					$k = 3$				
$p+\delta$ \ r	1	2	3	4	5	1	2	3	4	5
					$\delta = .00$					
.50	.933	.886	.856	.846	.842	.988	.945	.915	.904	.887
.75	.988	.970	.952	.936	.923	.999	.996	.991	.984	.977
.95	1.000	1.000	.999	.998	.997	1.000	1.000	1.000	1.000	1.000
1.00	1.000	1.000	1.000	1.000	1.000	1.000	1.000	1.000	1.000	1.000
					$\delta = .10$					
.50	.912	.858	.824	.811	.806	.975	.913	.875	.865	.840
.75	.979	.950	.923	.902	.885	.999	.998	.974	.959	.944
.95	.998	.994	.988	.980	.970	1.000	1.000	.999	.998	.997
1.00	.999	.998	.996	.992	.987	1.000	1.000	1.000	1.000	.999
					$\delta = .25$					
.50	.849	.764	.709	.679	.655	.918	.804	.737	.700	.646
.75	.948	.888	.837	.799	.768	.991	.947	.903	.866	.827
.95	.987	.962	.928	.891	.853	1.000	.993	.982	.966	.944
1.00	.992	.975	.948	.915	.878	.999	.997	.991	.982	.967
					$\delta = .50$					
.50	.500	.500	.500	.500	.500	.333	.333	.333	.333	.333
.75	.821	.691	.613	.570	.545	.900	.722	.599	.525	.458
.95	.920	.816	.724	.654	.604	.982	.898	.804	.709	.623
1.00	.937	.844	.750	.672	.613	.990	.927	.849	.758	.667

(*Contd.*)

Table 10.2 (*Contd.*)

$p^* = .90$

$p+\delta$ \ r	$k=2$: 1	2	3	4	5	$k=3$: 1	2	3	4	5
					$\delta = .00$					
.50	.998	.990	.984	.976	.973	.999	.998	.995	.991	.989
.75	.999	1.000	.999	.999	.997	.999	1.000	1.000	1.000	.999
.95	1.000	1.000	1.000	1.000	1.000	1.000	1.000	1.000	1.000	1.000
1.00	1.000	1.000	1.000	1.000	1.000	1.000	1.000	1.000	1.000	1.000
					$\delta = .10$					
.50	.995	.981	.970	.957	.951	.999	.994	.985	.979	.974
.75	1.000	.999	.997	.995	.991	1.000	1.000	.999	.999	.998
.95	1.000	1.000	1.000	1.000	1.000	1.000	1.000	1.000	1.000	1.000
1.00	1.000	1.000	1.000	1.000	1.000	1.000	1.000	1.000	1.000	.999
					$\delta = .25$					
.50	.973	.925	.885	.845	.813	.997	.960	.918	.885	.852
.75	.998	.990	.978	.959	.942	.999	.998	.993	.986	.978
.95	1.000	.999	.998	.996	.991	1.000	1.000	1.000	1.000	.999
1.00	1.000	1.000	.999	.998	.996	.999	1.000	1.000	1.000	.999
					$\delta = .50$					
.50	.500	.500	.500	.500	.500	.333	.333	.333	.333	.333
.75	.962	.871	.787	.707	.654	.996	.930	.834	.748	.671
.95	.996	.971	.933	.880	.823	.999	.995	.975	.946	.905
1.00	.998	.982	.955	.914	.863	.999	.998	.987	.969	.940

(*Contd.*)

Table 10.2 (Contd.)

$p^* = .95$

$p+\delta$ \ r	$k=2$: 1	2	3	4	5	$k=3$: 1	2	3	4	5
					$\delta = .00$					
.50	.999	.999	.997	.996	.994	.999	.999	.999	.999	.998
.75	.999	1.000	1.000	1.000	.999	.999	1.000	1.000	1.000	.999
.95	1.000	1.000	1.000	1.000	1.000	1.000	1.000	1.000	1.000	1.000
1.00	1.000	1.000	1.000	1.000	1.000	1.000	1.000	1.000	1.000	1.000
					$\delta = .10$					
.50	1.000	.997	.993	.999	.985	.999	.999	.998	.995	.993
.75	1.000	1.000	1.000	.999	.999	1.000	1.000	.999	1.000	1.000
.95	1.000	1.000	1.000	1.000	1.000	1.000	1.000	1.000	1.000	1.000
1.00	1.000	1.000	1.000	1.000	1.000	1.000	1.000	1.000	1.000	.999
					$\delta = .25$					
.50	.996	.975	.950	.925	.896	.999	.991	.974	.952	.929
.75	1.000	.999	.996	.991	.985	.999	.999	.999	.998	.996
.95	1.000	1.000	1.000	1.000	.999	1.000	1.000	1.000	1.000	1.000
1.00	1.000	1.000	1.000	1.000	1.000	.999	1.000	1.000	1.000	.999
					$\delta = .50$					
.50	.500	.500	.500	.500	.500	.333	.333	.333	.333	.333
.75	.998	.951	.885	.822	.754	.999	.984	.938	.871	.799
.95	1.000	.996	.982	.961	.929	.999	.999	.998	.990	.977
1.00	1.000	.998	.990	.977	.955	.999	.999	.999	.996	.990

where M is the smallest integer such that

$$\sum_{x=0}^{M} \binom{x+r-1}{x} p^r q^x \geqslant 0.999 \; (p > 0).$$

It appears from our tables that $E(S)$ is an increasing function of p, for fixed $p + \delta = p_k$ and an increasing function of p for fixed δ. This is very tempting for one to conjecture that, for $1 \leqslant m \leqslant k - 1$, $E(S; p_1 = p_2 = \ldots = p_m = p)$ is strictly increasing in p, though again a proof seems very difficult. Gupta (1965) has proven such a theorem for the case when the parameter of interest is one of location or scale and the distribution of the random variable used in the procedure has monotone likelihood ratio in x. The theorem does not always hold if the parameter is not one of location or scale, since for example it does not hold for the binomial distribution, which has monotone likelihood ratio. The proofs for establishing that $P\{CS\}$ and $E(S)$ are increasing functions probably depend on the same device.

When one compares our tables with Table 3 of Gupta and Sobel (1960), which gives $k^{-1}E(S)$ for the binomial case where one takes a sample of size n from each population and retains those populations having the most successes, one finds that, for given k and P^*, the negative binomial procedure sometimes gives a smaller, and at other times larger $E(S)$ than the binomial procedure, depending on the values of p and δ, when one looks at comparable values of r and n. The negative binomial procedure appears more favourable than the binomial one for smaller values of δ and larger values of p. In trying to decide which procedure to use, one should look at the expected number n of trials which will be necessary to achieve r successes, and compare $E(S)/k$ for a sample of size r using the negative binomial procedure with $E(S)/k$ using a binomial procedure with sample size n, at the parameter configuration of interest.

Example 5.1(i) Let $P^* = .75$, $k = 2$, and suppose we are interested in the expected proportion of populations retained for a configuration (5.3) with $p = .4$, $S = .1$. If we use the negative binomial procedure with $r = 4$, then we find from Table 10.2 that $E(S)/2 = .810$. The average number of items which will be tested, when $r = 4$, is $r + rq/p = 10$ items. Consulting Table 3 of Gupta and Sobel (1960) for the given values of k, P^*, p and δ, and for a sample size of $n = 20$, we find that for the binomial procedure, $E(S)/2 = .846$, thus the negative binomial procedure is preferred in this case, since it has the smaller value of $E(S)$.

(ii) For the same values of P^* and k, consider the configuration (5.3) with $p = .25$, $\delta = \cdot 50$. For $r = 5$, from Table 10.2, we find $E(S)/2 = .545$ for the negative binomial procedure. The average number of items we would expect to test would be $r + rq/p = 20$ items, and for $n = 20$ the binomial procedure has $E(S)/2 = .502$, as shown in Table 3 of Gupta and Sobel (1960). Thus in this case the binomial procedure is preferred.

10.6 OTHER PROPERTIES OF THE PROCEDURE R

(i) The Property of Monotonicity

Theorem 7.1. For $p_i \geqslant p_j$,

P (including population with p_i in the subset) $\geqslant$
P (including population with p_j in the subset).

Proof. We can write the rule (3.3) defining the procedure as "retain population π_i in the subset if and only if

$$W_i \leqslant b \min_j W_j, \tag{6.1}$$

where $W = X + r$ is the number of trials necessary to achieve r successes. The negative binomial distribution has monotone likelihood ratio in W, and thus the cdf's $F(y; r, q)$, $q = 1 - p$ are a stochastically increasing family, thus they satisfy the hypothesis of Theorem 2 of Rizvi (1964). The conclusion of that theorem is, in part, that if we use a rule of the form (6.1), for the problem of selecting a subset containing the population with the smallest θ-value, then if $\theta_i \leqslant \theta_j$,

P(including population with θ_i in subset) $\geqslant$
P(including population with θ_j in subset).

With $\theta = q = 1 - p$, we thus obtain the desired result.

(ii) The Property of Unbiasedness

The probability of rejecting any population not having the largest p-value is not less than the probability of rejecting the best population. In this sense, we say that the procedure is unbiased.

(iii) The probability of including the best population in the subset given the size S of the subset is maximum among the $\binom{k}{S}$ possible subsets of size S.

10.7 ASYMPTOTIC CONSIDERATIONS

In this section we consider the case when r is sufficiently large. Then appealing to the asymptotic normality of the x_i we get much simpler results. Since the probability of correct classification achieves its minimum when $p_1 = \ldots = p_m = p$.

$$P(\text{CS}) = P(X_k \leqslant bX_j + (b-1)r, j = 1, \ldots, k-1).$$

Now using the fact that $X_i \overset{d}{\approx}$ normal $(rq/p_1, rq/p_2)$, we let $Z_i = (pX_i - rq)/\sqrt{rq}$ and note that

$$\inf P(\text{CS}) = P\left(Z_i + \sqrt{rq} \leqslant b(Z_j + \sqrt{rq}) + \frac{(b-1)pr}{\sqrt{rq}}, j = 1, \ldots, k-1\right)$$

$$= \int_{-\infty}^{\infty} \Phi^{k-1}\left\{\left(\frac{b-1}{b}\right)\left(\sqrt{rq} + \frac{pr}{\sqrt{rq}}\right) - \frac{x}{b}\right\} d\Phi(x)$$

$$= \int_{-\infty}^{\infty} \Phi^{k-1}\left\{\left(\frac{b-1}{b}\right)\left(\frac{r}{q}\right)^{1/2} - \frac{x}{b}\right\} d\Phi(x). \tag{7.1}$$

The right side expression in (7.1) achieves its minimum when $q = 1$. Hence

$$\inf P(\text{CS}) = \int_{-\infty}^{\infty} \Phi^{k-1}\left\{\frac{(b-1)r^{1/2}-x}{b}\right\} d\Phi(x). \tag{7.2}$$

Gupta (1963) has tabulated values of the integral of the form

$$\int_{-\infty}^{\infty} \Phi^{N}\left(\frac{-x\rho^{1/2}+H}{(1-\rho)^{1/2}}\right) d\Phi(x)$$

for $N = 1, 2, \ldots, 12$ and $H = -3.50(.10)3.50$ and several values of ρ. Using these tables we can tabulate values of (7.2) by setting

$$\rho = (1+b^2)^{-1} \text{ and } H = \left(\frac{b-1}{b}\right)\sqrt{r}(1+b^{-2})^{-1/2}. \tag{7.3}$$

Expected Size of the Subset

Let

$$P_i(k, r, b, \tilde{p}) = P(X_i \leqslant bX_j + (b-1)r, \text{ for all } j \neq i)$$

$$= P\left(\frac{Z_i + \sqrt{rq_i} + rq_i}{p_i}\right) \leqslant b(Z_j\sqrt{rq_j} + rq_j) + (b-1)r, j \neq i)$$

$$= \int_{-\infty}^{\infty} \prod_{j=1, j\neq i}^{k} \Phi\left\{\frac{p_j[\sqrt{r}(q_jb - q_i) - x\sqrt{q_i}]}{p_i\sqrt{q_j}}\right\} d\Phi(x) \tag{7.4}$$

and

$$\text{Expected Size} = \sum_{i=1}^{k} P_i(k, r, b, \tilde{p}).$$

For the configuration $p_i = p$, for $i = 1, \ldots, k-1$ and $p_k = p + \delta$ we have

$$\text{Expected Size} = \int_{-\infty}^{\infty} \Phi^{k-1}\left[\frac{p\{\sqrt{r}(bq - q + \delta) - x\sqrt{q-\delta}\}}{(p+\delta)\sqrt{q}}\right] d\Phi(x)$$
$$+ (k-1)\int_{-\infty}^{\infty} \Phi^{k-2}\left[\frac{p\{\sqrt{r}(b-1)q - x\sqrt{q}\}}{p\sqrt{q}}\right]$$
$$\times \Phi\left[\frac{(p+q)\{\sqrt{r}(qb - \delta b - q) - x\sqrt{q}\}}{p\sqrt{q-\delta}}\right] d\Phi(x). \tag{7.5}$$

Remark 7.1. Using Theorem 9.3.1 or Theorem 12.3.7 of Wilks (1962) concerning variance stabilizing transformations, one can show that for large r, $\log[W + \{W(W-1)\}^{1/2} - 1/2]$ is asymptotically normally distributed with mean equal to $\log[p^{-1}(1+q^{1/2}) - 1/2]$ and variance equal to r^{-1} where $W = (X+r)/r$, that is the inverse of the maximum likelihood estimate of p. Using this result will also lead to the same results as above.

Acknowledgement: The authors thank the referee for a careful reading of the manuscript and for pointing out the use of Stirling numbers in the Lemma of the Appendix.

APPENDIX

In order to prove Theorem 4.1, we need the following lemma.

Lemma. For any positive integer s, and for all x, $0 \leqslant x \leqslant 1$, we have

(i) $\sum_{n=0}^{\infty} n^s x^n = P_s(x)(1-x)^{-(s+1)}$

where $P_s(x)$ is a polynomial of degree s in x, and if

$$P_s(x) = \sum_{t=0}^{s} a_t(s)x^t,$$

then (ii)

$$a_t(s) = \begin{cases} 0, & t = 0 \text{ or } t > s \\ ta_t(s-1) + (s-t+1)a_{t-1}(s-1), & 1 \leqslant t \leqslant s-1 \\ 1, & t = s \end{cases}$$

(iii) $P_s(1) \sum_{t=0}^{s} a_t(s) = s!$

(iv) $P_s'(1) \sum_{t=0}^{s} ta_t(s) = (s+1)!\big/2.$

Proof. The proof is by induction on s. For $s = 0$, $P_0(x) = 1$, and for $s = 1$, $\sum_{n=0}^{\infty} nx^n = x(1-x)^{-2}$ so that with $P_1(x) = x$, (i)–(iv) hold.

Now assume the truth of the propositions for $s - 1$. Then

$$\begin{aligned}
\sum_{n=0}^{\infty} n^s x^n &= \left\{x \frac{d}{dx}\right\} \sum_{n=0}^{\infty} n^{s-1}x^n \\
&= x \frac{d}{dx} \{P_{s-1}(x)(1-x)^{-s}\} \\
&= [x(1-x)P'_{-1}(x) + xsP_{s-1}(x)](1-x)^{-(s+1)} \\
&= \left[\sum_{t=0}^{s-1} \{t(x^t - x^{t+1}) + sx^{t+1}\}a_t(s-1)\right](1-x)^{-(s+1)} \\
&= \left[\sum_{t=0}^{s-1} \{ta_t(s-1)+(s-t+1)a_{t-1}(s-1)\}x^t + a_{s-1}(s-1)x^s\right](1-x)^{-(s+1)} \\
&= P_s(x)(1-x)^{-(s+1)},
\end{aligned}$$

where the coefficient of $P_s(x)$ clearly satisfies (ii). We have shown that an equivalent expression for $P_s(x)$ is

$$P_s(x) = \sum_{t=0}^{s-1} \{t(x^t - x^{t+1}) + sx^{t+1}\}a_t(s-1),$$

so that setting $x = 1$ and using the induction hypothesis

$$P_s(1) = s \sum_{t=0}^{s-1} a_t(s-1) = sP_{s-1}(1) = s\cdot(s-1)! = s!$$

proving (iii). Differentiating the above expression with respect to x, and setting $x = 1$, we obtain

$$\begin{aligned}
P_s'(1) &= \sum_{t=0}^{s-1} [t\{t-(t+1)\} + s(t+1)]a_t(s-1) \\
&= s\sum_{t=0}^{s-1} a_t(s-1) + (s-1)\sum_{t=0}^{s-1} ta_t(s-1) \\
&= sP_{s-1}(1) + (s-1)P_{s-1}'(1) \\
&= s(s-1)! + \tfrac{1}{2}(s-1)s! = (s+1)!/2
\end{aligned}$$

using the induction hypothesis. This completes the proof of (iv) and of the lemma.

Remark. Recall that

$$n^s = \sum_{j=0}^{s} A_{s,j} n_{(j)}$$

where $n_{(j)} = n(n-1)\ldots(n-j+1)$ and $A_{s,j}$ are the Stirling's numbers of the first kind. Hence

$$\begin{aligned}
\sum_{n=0}^{\infty} n^s x^n &= \sum_{j=0}^{s} A_{s,j} \sum_{n=0}^{\infty} n_{(j)} x^n \\
&= \sum_{j=0}^{s} A_{s,j} x^j \left[\frac{d^j}{dx^j} \sum_{n=0}^{\infty} x^n\right] \\
&= \sum_{j=0}^{s} A_{s,j} x^j (1-x)^{-(j+1)} \cdot j!.
\end{aligned}$$

So if

$$\sum_{n=0}^{\infty} n^s x^n = P_s(x)(1-x)^{-(s+1)},$$

then

$$\begin{aligned}
P_s(x) &= \sum_{j=0}^{s} A_{s,j} \cdot j! x^j (1-x)^{s-j} \\
&= \sum_{j=0}^{s} A_{s,j} j! \sum_{l=0}^{s-j} \binom{s-j}{l} (-1)^l x^{l+j} \\
&= \sum_{j=0}^{s} A_{s,j} j! \sum_{k=j}^{s} \binom{s-j}{k-j} (-1)^{k-j} x^k \\
&= \sum_{k=0}^{s} (-1)^k x^k \left\{\sum_{j=0}^{k} A_{s,j} j! \binom{s-j}{k-j} (-1)^j\right\}.
\end{aligned}$$

Thus, one can get an explicit expression for $P_s(x)$ in terms of Stirling numbers of the first kind where

$$A_{s,j} = \frac{1}{j!} \sum_{l=0}^{j} (-1)^{j-l} \binom{j}{l} l^s.$$

Proof of Theorem 4.1. Let $c = (b-1)r$. By tabulating some of the values of $\boldsymbol{x}$ and $f(x)$, one finds that (4.2) can be written as

$$P\{\text{CS}\}=\sum_{x=0}^{c}\binom{x+r-1}{x}p^r q^x+\sum_{x=0}^{\infty}\sum_{j=1}^{b}\binom{c+xb+j+r-1}{c+xb+j}p^r q^{c+xb+j}$$
$$\times\left[\sum_{y-x+1}^{\infty}\binom{y+r-1}{y}p^r q^y\right]^{k-1}, \tag{A.1}$$

and making use of a well-known identity involving the positive binomial and negative binomial cdf's, (A.1) can be rewritten as

$$P\{\text{CS}\}=F(c;\,r,\,p)+\sum_{j=1}^{b}\sum_{x=0}^{\infty}\binom{c+xb+j+r-1}{c+xb+j}p^r q^{c+xb+j}$$
$$\times\left[\sum_{y=0}^{r-1}\binom{x+r}{y}p^y q^{x+r-y}\right]^{k-1}.$$

Taking the $(k-1)$-th power of the term in brackets, one obtains

$$P\{\text{CS}\}=F(c;\,r,\,p)+\sum_{j=1}^{b}\sum_{x=0}^{\infty}\binom{c+xb+j+r-1}{c+xb+j}p^r q^{c+xb+j}\sum_{y_1=0}^{r-1}\ldots\sum_{y_{k-1}=0}^{r-1}$$
$$\left\{\prod_{i=1}^{k-1}\binom{x+r}{y_i}\right\}p^{\Sigma y_i}q^{(k-1)(x+r)-\Sigma y_i}$$
$$=F(c;\,r,\,p)+\sum_{j=1}^{b}\sum_{y_1=0}^{r-1}\ldots\sum_{y_{k-1}=0}^{r-1}\frac{p^{r+\Sigma y_i}q^{c+j+(k-1)r-\Sigma y_i}}{(r-1)!\prod_{i=1}^{k-1}y_i!}$$
$$\times\sum_{x=0}^{\infty}\frac{(c+xb+j+r-1)!}{(c+xb+j)!}\left\{\prod_{i=1}^{k-1}\frac{(x+r)!}{(x+r-y_i)!}\right\}q^{(b+k-1)x},$$

where,

$$G(x,\,b,\,j,\,r,\,\tilde{y})=\frac{(c+xb+j+r-1)!\{(x+r)!\}^{k-1}}{(c+xb+j)!\prod_{i=1}^{k-1}(x+r-y_i)!}$$

is a polynomial of degree $r-1+\sum_{i=1}^{k-1}y_i$ in x, the coefficients of which are positive integers depending on b, j, r, and $\tilde{y}=(y_1,\ldots,y_{k-1})$. Writing

$$G(x,\,b,\,j,\,r,\,\tilde{y})=\sum_{s=0}^{r-1+\Sigma y_i}g(s,\,b,\,j,\,r,\,\tilde{y})x^s,$$

one obtains

$$P\{\text{CS}\}=F(c;\,r,\,p)+\sum_{j=1}^{b}\sum_{y_1=0}^{r-1}\ldots\sum_{y_{k-1}=0}^{r-1}\frac{p^{r+\Sigma y_i}q^{c+j+(k-1)r-\Sigma y_i}}{(r-1)!\prod_{i=1}^{k-1}y_i!}$$
$$\times\sum_{s=0}^{r-1+\Sigma y_i}g(s,\,b,\,j,\,r,\,\tilde{y})\sum_{x=0}^{\infty}x^s q^{(b+k-1)x},$$

and applying (i) of the lemma, one obtains (4.4). Let

$$h(s,\,p)=\frac{p^{r+\Sigma y_i}}{[1-q^{b+k-1}]^{s+1}}=\frac{p^{r-1+\Sigma y_i-s}}{[1+q+\ldots+q^{b+k-2}]^{s+1}}.$$

Then,

$$\lim_{p\to 0} h(s, p) = \begin{cases} 0, & s<r-1+\Sigma y_i \\ (b+k-1)^{-(r+\Sigma y_i)}, & s=r-1+\Sigma y_i \text{ where } t=(b+k-1)^{-1}. \end{cases}$$

The coefficients of $x^{r-1+\Sigma y_i}$ in $G(x, b, j, r, \tilde{y})$ is b^{r-1}, and by (iii) of the lemma, $P_{r-1+\Sigma y_i}(1) = (r-1+\Sigma y_i)!$, thus

$$\lim_{p\to 0} P\{\text{CS}; k, r, b, p\}$$

$$= \sum_{j=1}^{b} \sum_{y_1=0}^{r-1} \dots \sum_{y_{k-1}=0}^{r-1} \frac{b^{r-1}(r-1+\Sigma y_i)!}{(r-1)! \sum_{i=1}^{k-1} y_i!(b+k-1)^{r+\Sigma y_i}}$$

$$= \sum_{y_1=0}^{r-1} \dots \sum_{y_{k-1}=0}^{r-1} \frac{(r-1+\Sigma y_i)!}{(r-1)! \prod_{i=1}^{k-1} y_i!} \left(\frac{b}{b+k-1}\right)^r \left(\frac{1}{b+k-1}\right)^{\Sigma y_i}$$

which completes the proof of (4.5). Performing the required differentiation, and using (iii) and (iv) of the lemma, one obtains, for $r > 1$,

$$\lim_{p\to 0} \frac{d}{dp}\left[\frac{p^{r-1+\Sigma y_i-s} q^{c+j+(k-1)r-\Sigma y_i} P_s(q^{b+k-1})}{(1+q+\dots+q^{b+k-2})^{s+1}}\right]$$

$$= \begin{cases} 0, \; s < r-2+\Sigma y_i \\ \dfrac{(r-2+\Sigma y_i)!}{(b+k-1)^{r-1+\Sigma y_i}}, \; s = r-2+\Sigma y_i \\ \dfrac{(r-1+\Sigma y_i)!}{(b+k-1)^{r+\Sigma y_i}} [\tfrac{1}{2}(r+\Sigma y_i) - c - j - kr], \; s = r-1+\Sigma y_i \end{cases}$$

The coefficient of $x^{r-2+\Sigma y_i}$ in $G(x, b, j, r, \tilde{y})$ is

$$b^{r-2}(r-1)(c+j+\tfrac{1}{2}r) + b^{r-1} \sum_{i=1}^{k-1} \tfrac{1}{2} y_i(2r - y_i + 1)$$

as can be verified by partially expanding G. Combining the above results, one obtains, for $r > 1$,

$$\lim_{p\to 0} \frac{d}{dp} P\{\text{CS}\} = \sum_{j=1}^{b} \sum_{y_1=0}^{r-1} \dots \sum_{y_{k-1}=0}^{r-1} \frac{b^{r-2}(r-2+\Sigma y_i)!}{(r-1)! \prod_{i-1}^{k-1} y_i!\,(b+k-1)^{r+\Sigma y_i}}$$

$$\times [(b+k-1)\{(r-1)(c+j+\tfrac{1}{2}r) + \tfrac{1}{2}b \sum_{i=1}^{k-1} y_i(2r-y_i+1)\}$$

$$+ b(r-1+\Sigma y_i)\{\tfrac{1}{2}(r+\Sigma y_i) - c - j - kr\}].$$

After regrouping terms, recalling that $c = (b-1)r$, the expression in brackets, when summed over j, becomes

$$\tfrac{1}{2}(k-1)(r-1)(b+1-r)$$

$$+ \tfrac{1}{2}b^2(2r-b-1) \sum_{i=1}^{k-1} y_i - \tfrac{1}{2}b^2(b+k-2) \sum_{i=1}^{k-1} y_i(y_i-1) + \tfrac{1}{2}b^2 \sum_{i\neq l}\sum y_i y_l$$

Because of the symmetry of the above expression in $y_1, \ldots, y_{k-1}$, we can replace Σy_i by $(k-1)y_1$, $\Sigma y_i(y_i-1)$ by $(k-1)y_1(y_1-1)$, and $\Sigma\Sigma y_i y_l$ by $(k-1)(k-2)y_1 y_2$, with the result that, since terms with $y_1 = 0$ or $y_2 = 0$ drop out,

$$\lim_{p\to 0} \frac{d}{dp} P\{\text{CS}\}$$

$$= \frac{k-1}{2(b+k-1)} \Bigg[(b+1-r) \sum_{y_1=0}^{r-1} \sum_{y_2=0}^{r-1} \cdots \sum_{y_{k-1}=0}^{r-1} \frac{(r-2+\Sigma y_i)!}{(r-2)! \prod_{i=1}^{k-1} y_i!} \left(\frac{b}{b+k-1}\right)^{r-1} \left(\frac{1}{b+k-1}\right)^{\Sigma y_i}$$

$$+ (2r-b-1) \sum_{y_1=1}^{r-1} \sum_{y_2=0}^{r-1} \cdots \sum_{y_{k-1}=0}^{r-1} \frac{(r-2+\Sigma y_i)!}{(r-1)!\,(y_1-1)! \prod_{i=2}^{k-1} y_i!} \left(\frac{b}{b+k-1}\right)^{r} \left(\frac{1}{b+k-1}\right)^{\Sigma y_i - 1}$$

$$- \frac{r(b+k-2)}{b} \sum_{y_1=2}^{r-1} \sum_{y_2=0}^{r-1} \cdots \sum_{y_{k-1}=0}^{r-1} \frac{(r-2+\Sigma y_i)!}{r!(y_1-2)! \prod_{i=2}^{k-1} y_i!} \left(\frac{b}{b+k-1}\right)^{r+1} \left(\frac{1}{b+k-1}\right)^{\Sigma y_i - 2}$$

$$+ \frac{r(k-2)}{b} \sum_{y_1=1}^{r-1} \sum_{y_2=1}^{r-1} \cdots \sum_{y_{k-1}=0}^{r-1} \frac{(r-2+\Sigma y_i)!}{r!(y_1-1)!\,(y_2-1)! \prod_{i=3}^{k-1} y_i!} \left(\frac{b}{b+k-1}\right)^{r+1}$$

$$\times \left(\frac{1}{b+k-1}\right)^{\Sigma y_i - 2} \Bigg]$$

which, after making several obvious changes of variable, reduces to (4.6). Setting $r = 1$ in (4.4), we have

$$P\{\text{CS};\, k,\, 1,\, b,\, p\} = \sum_{x=0}^{b-1} pq^x + \sum_{j=1}^{b} \frac{pq^{b+k-2+j}}{1 - q^{b+k-1}}.$$

After performing the summations and combining terms, one obtains

$$P\{\text{CS};\, k,\, 1,\, b,\, p\} = \frac{1 - q^b}{1 - q^{b+k-1}}, \tag{A.2}$$

so that an application of L'Hôpital's rule gives

$$\lim_{p\to 0} P\{\text{CS};\, k,\, 1,\, b,\, p\} = \frac{b}{b+k-1}, \tag{A.3}$$

and, by straightforward differentiation of (A.2), we obtain

$$\lim_{p\to 0} \frac{d}{dp} P\{\text{CS};\, k,\, 1,\, b,\, p\} = \frac{b(k-1)}{2(b+k-1)},$$

which we note is positive for all $b \geqslant 1$ and $k \geqslant 2$. Thus, we have verified (4.5) and (4.6) for $r = 1$. Thus, completes the proof of Theorem 4.1.

We now take a heuristic approach in order to argue that (4.5) remains at least approximately true when b is not an integer. Similar to (A.1) we have, for general b,

$$P\{\text{CS};\, k, r, b, p\} = \sum_{x=0}^{[c]} \binom{x+r-1}{x} p^r q^x + \sum_{x=0}^{\infty} \sum_{j=1}^{m(x)} \binom{[c+xb]+j+r-1}{[c+xb]+j}$$

$$\times p^r q^{[c+xb]+j} \left[\sum_{y=x+1}^{\infty} \binom{y+r-1}{y} p^r q^y \right]^{k-1} \quad \text{(A.4)}$$

where $c = (b-1)r$ and $m(x)$ is either $[b]$ or $[b]+1$. Roughly speaking, the average value of $m(x)$, over all x, is b, and the average value of $[c+xb]+j$ is $[c]+xb$. Hence, let us approximate (A.4) by

$$P\{\text{CS};\, k, r, b, p\} \cong \sum_{x=0}^{[c]} \binom{x+r-1}{x} p^r q^x + b \sum_{x=0}^{\infty} \binom{[c]+xb+r-1}{[c]+xb} p^r q^{[c]+xb}$$

$$\times \left[\sum_{y=x+1}^{\infty} \binom{y+r-1}{y} p^r q^y \right]^{k-1}. \quad \text{(A.5)}$$

A careful analysis of the proof of Theorem 4.1 shows that after the derivation of A.1, no use is made of the assumption that b is an integer, and also that, considering $(c+xb+j+r-1)!/(c+xb+j)!$ to be a polynomial of degree $r-1$ in x, only the term in x^{r-1}, when summed over x, contributes a non-zero term to $P\{\text{CS}\}$ as p approaches zero. The coefficient of x^{r-1} is b^{r-1}, and does not depend on c or j. Thus, had we replaced $c = (b-1)r$ by some other integer and the summation on j by the coefficient b in A.1, we would still have obtained (4.5), that is, the right hand side of A.5 also approaches (4.5) as p approaches zero.

Since $b(k, r, p^*)$ is to be determined as the solution of (4.3), we conclude from Theorem 4.1 and the above remarks that $b(k, r, p^*)$ is approximately given by the solution of

$$F(r-1, \ldots, r-1;\, r, bt, t, \ldots, t) = P^*. \quad \text{(A.6)}$$

REFERENCES

1. Bartlett, N. S. (1965): *Ranking and Selection Procedures for Continuous and Certain Discrete Distributions*, Ph.D. thesis, Case Institute of Technology, Cleveland, Ohio.
2. Govindarajulu, Z. and Hubacker, N. W. (1964): Percentiles of order statistics in samples from uniform, normal, chi (1 d.f.) and Weibull populations. *Rep. Statist. Appl. Res.*, JUSE **11**, 64–90.
3. Gupta S. S. (1956): On a decision rule for a problem in ranking means. Mimeo Series No. **150**, *Inst. of Statist.*, Univ. of North Carolina.
4. ——— (1963): Probability integrals of multivariate normal and multivariate *t*. *Ann. Math. Statist.*, **34**, 792–828.
5. ——— (1965): On some multiple decision (selection and ranking) rules, *Technometrics*, **6**, 225–245.
6. Gupta, S. S. and Sobel, M. (1960): *Selecting a subset containing the best of several binomial populations*, Chapter 20 in Olkin, I. (ed.) *Contributions to Probability and Statistics*, Stanford Univ. Press.
7. Pearson, E. S. and Hartley, H. O. (eds.) (1956): *Biometrika Tables for Statistics*, Volume **1**, Cambridge Univ. Press, London.

8. Pearson, K. (ed.) (1934): *Tables of the Incomplete Beta Function*, Cambridge Univ. Press, London.

9. Rizvi, M. H. (1964): A note on monotonicity properties of a class of ranking and selection procedures, Abstract, *Ann. Math. Statist*, **35**, 930–931.

10. Sobel, M. and Huyett, M. J. (1957): Selecting the best one of several binomial populations. *Bell Syst. Tech. J.*, **36**, 537–576.

11. Wilks, S. S. (1962): *Mathematical Statistics*, Wiley, New York.

12. Guttman, I. and Tiao, G. C. (1964): A Bayesian approach to some best population problems, *Ann. Math. Statist.*, **35**, 825–835.

13. Gupta S. S. and Nagel, K. (1971): On some contributions to multiple decision theory, *Statistical Decision Theory and Related Topics* (Ed. S. S. Gupta and J. Yackel), Academic Press, New York, pp. 79–102.

11

Sequential Fixed Size Confidence Regions for the Mean Vector in the Bivariate Normal with known Covariance Matrix

M. S. Prasad and **Alladi Subrahmanyam**

Department of Statistics
Shivaji University
Kolhapur-416 004, India

Department of Mathematics
Indian Institute of Technology
Powai, Bombay, India

11.1 INTRODUCTION

Let $(\Omega, \mathscr{F})$ be a measurable space and let $\mathbb{P}$ be a family of probability measures on $(\Omega, \mathscr{F})$. Let $\{X_n, n \geqslant 1\}$ be a sequence of random variables on $(\Omega, \mathscr{F})$ such that one of the probability measures in $\mathbb{P}$ governs the distributions associated with $\{X_n\}$. Let $\theta = \theta(P)$ be a functional from $\mathbf{P}$ to Θ, which is a subset of $\mathbf{R}^k$, the k-dimensional Euclidean space. The problem is that of obtaining 'fixed size' Confidence Regions for θ. Let $\mathbb{B}_k$ be the class of all Borel sets of $\mathbf{R}^k$.

A sequential procedure is the pair (R_N, N), where N is a stopping time with respect to the family of σ-fields $\{\sigma(X_1, \ldots, X_n), n \geqslant 1\}$ and $R_n \subset \mathbb{B}_k$ is the region to be used as the confidence region for θ, when $N = n$. In the pair (R_N, N), R_N is written as the first component to indicate that, first one should decide what type of confidence regions $\{R_n, n \geqslant 1\}$ are to be obtained, since the stopping time will depend on the 'shape' of the region desired. When N is degenerate at m (not depending on $P \in \mathbb{P}$) we say that the procedure is a fixed sample procedure.

Let $\mathcal{C}$ be a subclass of $\mathbb{B}_k$. *A* procedure (R_N, N) is said to be a *fixed size procedure* if R_n belongs to $\mathcal{C}$ for every n. For example, $\mathcal{C}$ may be taken to be the class of all symmetric convex sets of specified diameter $2d$ or as the class of all sets with specified volume d or as any other class in which the size is controlled by an appropriate characteristic.

We shall denote the subclass $\mathcal{C}$ by $\mathcal{C}(d)$, when $d(> 0)$ is the specified size. It may be noted that when $k = 1$, that is in the scalar parameter case, all the usual definitions of 'size' reduce to the lebesgue measure of the set. It is clear that the stopping time N and hence the procedure (R_N, N) depend on the class $\mathcal{C}(d)$ and so on d. We shall denote the procedure $(R_{N(d)}, N(d))$ by S_d.

A fixed size procedure S_d is said to be α-consistent if

$$P_Q\{R_N \ni \theta(Q)\} \geqslant \alpha \tag{1.1}$$

for all $Q \in \mathbf{P}$.

An α-consistent procedure S_d is said to be optimal (within the class of all α-consistent procedures) if the effective sample size

$$n_0(S_d) = \sup_{Q \in \mathbf{P}} E_{\theta(Q)}(N_d)$$

is minimum.

When $\{X_n, n \geqslant 1\}$ is an i.i.d. sequence of random variables having a normal distribution with mean θ and known variance σ^2, Stein and Wald (1947) have shown that the minimum effective sample size is given by

$$n_0(d) = \left[\frac{a_\alpha \sigma^2}{d^2}\right] + 1, \tag{1.2}$$

which gives a confidence interval of length $2d$, where a_α is chosen so that the procedure becomes α-consistent.

However, in general, an α-consistent procedure may not exist (cf. Zacks, 1971). So we consider asymptotically α-consistent procedures S, which is a family $\{S_d, d > 0\}$ of fixed size procedures. A procedure $S = \{S_d, d > 0\}$ is said to be asymptotically α-consistent if

$$\lim_{d \downarrow 0} P_Q\{R_{N(d)} \ni \theta(Q)\} \geqslant \alpha \tag{1.3}$$

for all $Q \in \mathbf{P}$.

When $\{X_n, n \geqslant 1\}$ is a sequence of i.i.d. random variable with mean θ, and (unknown) variance σ^2, Chow and Robbins (1965) have defined an asymptotically α-consistent procedure S to be asymptotically efficient if

$$\overline{\lim_{d \downarrow 0}} \frac{E_\theta[N(d)]}{n_0(d)} \leqslant 1 \tag{1.4}$$

for all $\theta \in \Theta$, where $n_0(d)$ is as defined in (1.2).

As pointed out by Zacks (1971) the motivation for the above definition of asymptotic optimality is provided by the theorem of Stein and Wald (1947).

The problem of constructing fixed size confidence regions, when $k \geqslant 2$, has been considered in the literature for certain families $\mathcal{C}(d)$. The optimality is defined through

$$\lim_{d \downarrow 0} \frac{E_\theta(N(d))}{n_0(d)}.$$

However, it is not clear how the numbers $n_0(d)$ were chosen and how the same should vary with $\mathcal{C}(d)$. In this paper, we answer this question.

In order to do this, we first prove a theorem analogous to the theorem of Stein and Wald (1947) when $k = 2$. This theorem is proved in section 11.3. This theorem is of independent interest and also motivates a definition of asymptotic optimality for fixed size confidence region. We also give a definition of optimality when $k > 3$, but with a weaker justification.

In section 11.4, examples are worked out to illustrate the particular cases of the unified theory. In section 11.5, a definition of the asymptotic efficiency of fixed size procedures based on CAN estimators is given. Our definition of optimality does not agree with that of Mukhopadhyaya (1981). In fact, his definition does not seem to be appropriate as shown by example 4.1.

11.2 THE PROBLEM AND THE PROPOSED SOLUTION

Let $\{X_n, n \geqslant 1\}$ be a sequence of bivariate vectors with bivariate normal distribution $N(\mu, \Sigma)$. The mean vector $\mu \in R^2$ is unknown and the covariance matrix $\boldsymbol{\Sigma}$ is a known 2×2 positive definite matrix. The largest and the smallest characteristic roots of Σ will be denoted by λ and λ_1 respectively. Henceforth $N(\mu, \Sigma)$ will denote either the random variable or the distribution, the meaning being clear from the context. Suppose we are interested in finding a fixed size region for μ belonging to the class $\mathcal{C} \subset \mathbb{B}_2$.

We make the following assumptions on the class $\mathcal{C}$.

Assumption 1. There exists a symmetric convex set C_0 in $\mathcal{C}$ such that

$$\sup_{C \in \mathcal{C}} \int_C dN(0, \Sigma) = \int_{C_0} dN(0, \Sigma) \tag{2.1}$$

Since C_0 is a symmetric convex set, there exists a Minkovski metric (cf. Rockafellar 1970, p. 132) $\rho(\cdot)$ such that

$$\sigma C_0 = \{y \in R^2 \mid \rho(y; 0) \leqslant \sigma\} \tag{2.2}$$

where $\sigma C_0 = \{\sigma x \mid x \in C_0\}$. Let $Z = \rho^2(X; 0)$, where X is a $N(0, \Sigma)$ vector and $\rho(\cdot)$ is the Minkovski metric corresponding to C_0.

Let $H(\cdot, \Sigma)$ denote the distribution function of Z, $f(x, \Sigma)$ the corresponding density, that is,

$$H(t, \Sigma) = \int_0^t f(x, \Sigma)\, dx. \tag{2.3}$$

Assumption 2. The function $f(x, \Sigma)$ in (2.3) is strictly decreasing in x.

For any procedure S, define the following two quantities

$$\alpha(S) = \inf_{\mu} P_\mu(R_N \text{ covers } \mu);\ \mu \in R^2 \tag{2.4}$$

and

$$n_0(S) = \sup_{\mu} E_\mu(N);\ \mu \in R^2 \tag{2.5}$$

We will call $\alpha(S)$ the coverage probability and $n_0(S)$ the effective sample size.

Denote by $S(\nu, c)$ the procedure (R_N, N) where $R_n = C_0 + Y_n$ and N is the stopping rule with distribution given by

$$P(N = n) = \begin{cases} 0 & \text{if } n \neq \nu, \nu + 1 \\ c & \text{if } n = \nu \\ 1 - c & \text{if } n = \nu + 1 \end{cases} \tag{2.6}$$

Here, $Y_n = \sum_{i=1}^{n} X_i/n$, the sample mean vector from n observations.

Let $\alpha[S(\nu, c)]$ denote the coverage probability of $S(\nu, c)$ and let $n_0[S(\nu, c)]$ denote the required sample size. Observe that

$$\begin{aligned} \alpha[S(\nu, c)] &= cP_\mu\{Y_\nu + C_0 \ni \mu\} + (1 - c)P_\mu\{Y_{\nu+1} + C_0 \ni \mu\} \\ &= cP_\mu\{\rho^2(Y_\nu - \mu; 0) \leqslant 1\} + (1 - c)P_\mu\{\rho^2(Y_{\nu+1} - \mu; 0) \leqslant 1\} \\ &= cH(\nu, \Sigma) + (1 - c)H(\nu + 1, \Sigma) \end{aligned} \tag{2.7}$$

Also, we see that

$$\begin{aligned} n_0[S(\nu, c)] &= c\nu + (1 - c)(\nu + 1) \\ &= \nu + 1 - c \\ &= \nu + a \end{aligned} \tag{2.8}$$

with $1 - c = a$.

In the next section, we will show that the procedure $S(\nu, c)$ is efficient according to the following definition.

Definition 2.1. A procedure S is said to be *efficient* if for any other fixed size procedure S' such that

$$\alpha(S) \leqslant \alpha(S')$$

we have

$$n_0(S) \leqslant n_0(S').$$

11.3. THE STEIN-WALD THEOREM IN R^2

In this section, we show that the procedure $S(\nu, c)$ is efficient. We first develop some notation. Let

$$\alpha(\mu, S) = P_\mu(R_N \text{ covers } \mu) \tag{3.1}$$

Assume that there is a $N(0, \epsilon^{-1}\Sigma)$ prior for μ. The symbol E^* denotes the expectation with respect to the joint distribution of $\{\mu, X_n, n \geqslant 1\}$. Define

$$\bar{\alpha}(\epsilon, S) = \int \alpha(\mu, S)\, dN(0, \epsilon^{-1}\Sigma) \tag{3.2}$$

$$\bar{E}(N, \epsilon, S) = \int E_\mu(N)\, dN(0, \epsilon^{-1}\Sigma).$$

In view of (2.4) and (2.5), it follows that

$$\bar{\alpha}(\epsilon, S) \geqslant \alpha(S) \quad \text{and} \quad \bar{E}(N \mid \epsilon, s) \leqslant n_0(S). \tag{3.3}$$

Let $\{p_m(\mu, S), m \geqslant 1\}$ denote the (conditional) probability mass function of N, the sample size required by the procedure S (given μ) and let

$$\bar{p}_m(\epsilon, S) = \int p_m(\mu, S)\, dN(0, \epsilon^{-1}\Sigma). \tag{3.4}$$

It follows from (3.2), (3.3) and the above that

$$\sum_{m=1}^{\infty} m\bar{p}_m(\epsilon, S) = \bar{E}(N, \epsilon, S) \leqslant n_0(S) \tag{3.5}$$

Also,

$$\begin{aligned}\bar{\alpha}(\epsilon, S) &= E^*(I[R_N \ni \mu]) \\ &= E^*\left(\sum_{m=1}^{\infty} I[N = m]\cdot I[R_m \ni \mu]\right) \\ &= \sum_{m=1}^{\infty} E^* I[N = m]\cdot E^*(I[R_m \ni \mu] \mid \mathbb{F}_m)\end{aligned} \tag{3.6}$$

where $I[A]$ denotes the indicator function of the set A and m denotes the σ-field generated by $X_1, \ldots, X_m$.

Since μ has a $N(0, \epsilon^{-1}\Sigma)$ prior, we see from the sufficiency of Y_m when m observations are made and the result in Berger (1980, page 157), that the conditional distribution of μ given $\mathbb{F}_m$ is $N(\mu_m, (m + \epsilon)^{-1}\Sigma)$ where $\mu_m = mY_m/(m + \epsilon)$. Using Assumption I, we get that

$$\begin{aligned}E^*(I[R_m \ni \mu] \mid \mathbb{F}_m) &\leqslant E^*(I[\mu_m + C_0 \ni \mu] \mid \mathbb{F}_m) \\ &= E^*(I[\rho^2(\mu - \mu_m; 0)] \leqslant 1 \mid \mathbb{F}_m) \\ &= H(m + \epsilon; \Sigma)\end{aligned}$$

Using this inequality in (3.6), we get

$$\bar{\alpha}(\epsilon, S) \leqslant \sum_{m=1}^{\infty} \bar{p}_m(\epsilon, S) H(m + \epsilon; \Sigma)$$

which alongwith (3.2) gives

$$\alpha(S) \leqslant \sum_{m=1}^{\infty} \bar{p}_m(\epsilon, S) H(m + \epsilon; \Sigma) \tag{3.7}$$

The following lemma will be used to prove the efficiency of $S(\nu, c)$.

Lemma 3.1. For any procedure S, if

$$\bar{E}(N; \epsilon, S) = \sum_{m=1}^{\infty} m\bar{p}_m(\epsilon, S) \leqslant \nu + a \tag{3.8}$$

where ν is a positive integer and $0 < a < 1$, then,

$$\sum_{m=1}^{\infty} \bar{p}_m(\epsilon, S) H(m + \epsilon; \Sigma) \leqslant (1 - a) H(\nu + \epsilon, \Sigma) + aH(\nu + 1 + \epsilon, \Sigma) \tag{3.9}$$

Proof. For the duration of the proof, we write $H(m+\epsilon)$ for $H(m+\epsilon, \Sigma)$, $f(t)$ for $f(t, \Sigma)$ and $\bar{p}_m$ for $\bar{p}_m(\epsilon, S)$.

Suppose that there exists a procedure S for which (3.8) is satisfied, but

$$\sum_{m=1}^{\infty} \bar{p}_m H(m+\epsilon) > \sum_{m=1}^{\infty} p_m^* H(m+\epsilon) \tag{3.10}$$

where p_m^* is the probability mass function of the stopping time corresponding to the procedure $S(\nu, c)$ note that

$$\sum_{m=1}^{\infty} m p_m^* = \nu + a.$$

From (3.8), (3.9), (3.10) and the above, we have that

$$0 < \sum_{m=1}^{\infty} (\bar{p}_m - p_m^*) H(m+\epsilon) - [H(\nu+1+\epsilon) - H(\nu+\epsilon)] \sum_{m=1}^{\infty} m(\bar{p}_m - p_m^*),$$

which may be rewritten as [because $\{\bar{p}_m, m \geqslant 1\}$ and $\{p_m^*, m \geqslant 1\}$ are probability mass functions]

$$0 < \sum_{m=1}^{\infty} \bar{p}_m - p_m^*) K_m \tag{3.11}$$

where,

$$K_m = H(m+\epsilon) - H(\nu+\epsilon) - (m-\nu)[H(\nu+1+\epsilon) - H(\nu+\epsilon)].$$

Note that $K_\nu = K_{\nu+1} = 0$. Now, we show that $K_m < 0$ for all $m \neq \nu$, $\nu + 1$.

If $m < \nu$, we have

$$\begin{aligned} K_m &= -\int_{m+\epsilon}^{\nu+\epsilon} f(t)\, dt + (\nu - m) \int_{\nu+\epsilon}^{\nu+1+\epsilon} f(t)\, dt \\ &< -(\nu - m) f(\nu+\epsilon) + (\nu - m) f(\nu+\epsilon) \\ &= 0. \end{aligned}$$

In the above, the inequality follows because f is strictly decreasing by Assumption 2. Similarly for $m > \nu + 1$, we have

$$\begin{aligned} K_m &= \int_{\nu+\epsilon}^{m+\epsilon} f(t) - (m - \nu) \int_{\nu+\epsilon}^{\nu+1+\epsilon} f(t)\, dt \\ &= \int_{\nu+1+\epsilon}^{m+\epsilon} f(t)\, dt - (m - \nu - 1) \int_{\nu+\epsilon}^{\nu+1+\epsilon} f(t)\, dt \\ &< (m - \nu - 1) f(\nu+1+\epsilon) - (m - \nu - 1) - f(\nu+1+\epsilon) \\ &= 0. \end{aligned}$$

This property of K_m's together with the fact that $p_m^* = 0$ for $m \neq \nu$, $\nu + 1$ implies that

$$\sum_{m=1}^{\infty} (\bar{p}_m - p_m^*) K_m < 0$$

which is a contradiction to (3.11). Therefore, we get that

$$\sum_{m=1}^{\infty} \bar{p}_m H(m + \epsilon) \leqslant \sum_{m=1}^{\infty} p_m^* H(m + \epsilon)$$
$$= (1 - a)H(\nu + \epsilon) + aH(\nu + 1 + \epsilon).$$

This proves the lemma.

Remark 3.1. If S is any procedure such that $n_0(S) \leqslant \nu + a$, where ν is a positive integer and $0 < a < 1$, then (3.5), (3.7) and Lemma 1 imply that for every $\epsilon > 0$

$$\alpha(S) \leqslant (1 - a)H(\nu + \epsilon, \Sigma) + aH(\nu + 1 + \epsilon, \Sigma).$$

Since H is a distribution function, by taking $\epsilon \to 0$ in the above, we get that

$$\alpha(S) \leqslant \alpha[S(\nu, c)]$$

where $S(\nu, c)$ is a procedure of the form (2.6) and $c = 1 - a$ as given above.

Theorem 3.1. The procedure $S(\nu, c)$ is efficient according to Definition (2.1).

Proof. If possible let there be a procedure S' such that

$$\alpha[S(\nu, c)] \leqslant \alpha(S') \tag{3.12}$$

and

$$n_0(S') < n_0[S(\nu, c)] \tag{3.13}$$

Let ν' be the integral part and a' be the fractional part of $n_0(S')$. Then $n_0(S') = \nu' + a'$. Let $S(\nu', c')$ (with $c' = 1 - a'$) be the procedure of the form (2.6). Then, from (3.13) it follows that

$$n_0(S') = n_0[S(\nu', c')] < n_0[S(\nu, c)].$$

Then, in view of the Remark 3.1

$$\alpha(S') \leqslant \alpha[S(\nu', c')]$$
$$< \alpha[S(\nu', c')].$$

The last inequality follows because $\alpha[S(\nu, c)]$ is a strictly increasing function of $n_0[S(\nu, c)]$. This is a contradiction to (3.12), hence this proves the theorem.

11.4 EXAMPLES

In this section we give some examples illustrating Theorem 3.1. We also make some remarks regarding the problem in higher dimensions. Throughout this section U and V will denote independent chi-square variates with one degree of freedom, $x = (x_1, x_2)$ and $y = (y_1, y_2)$ will denote points in $\mathbb{R}^2$. The largest and smallest characteristic roots of Σ are denoted by λ and λ_1 respectively. The symbol Z will denote the random variable $\rho^2(X; 0)$ where X is $N(0, \Sigma)$

Example 4.1. Let $\mathcal{C}$ be the class of all symmetric convex sets of diameter $2d$. We are interested in constructing confidence regions belonging to $\mathcal{C}$. The set C_0 which satisfies (2.1) is the circle of radius d centered at the origin. The Minkovski metric corresponding to C_0 is

$$\rho(x, y) = \{(x_1 - x_2)^2 + (y_1 - y_2)^2\}^{1/2}.$$

The random variable Z has the distribution of $\lambda U + \lambda_1 V$. Therefore

$$\begin{aligned} Pr\{\rho^2(X; 0) \leqslant x\} &= Pr\{U + \beta V \leqslant x/\lambda\} \\ &= H(x/\lambda, \Sigma) \end{aligned}$$

where $H(\cdot, \Sigma)$ is the distribution function of the random variable $U + \beta V$, $\beta = \lambda_1/\lambda$, where U and V are i.i.d. random variables having the chi-square distribution with one degree of freedom and $0 \leqslant \beta \leqslant 1$. In the following Lemma, we verify that the density function of $U + \beta V$ is strictly decreasing.

Lemma 4.1. The density function of $U + \beta V$ is strictly decreasing, for every $0 \leqslant \beta \leqslant 1$, where U and V are as described above.

Proof. The proof is trivial, when $\beta = 1$, since in this case $U + V$ has chi-square distribution with 2 degrees of freedom. Suppose that $0 \leqslant \beta < 1$. Let $F_\beta(x)$ denote Prob. $\{U + \beta V \leqslant x\}$. We then have

$$F_\beta(x) = k \int_0^{x/\beta} \text{Prob. } \{U \leqslant x - \beta v\} e^{-v/2} v^{-1/2}\, dv$$

where,

$$k = (2\pi)^{-1/2}.$$

Using the generalized form of Leibnitz's rule [cf. Theorem IV, page 324 of Ohmstead (1970)] we have

$$\begin{aligned} f_\beta(x) &= k^2 \int_0^{x/\beta} e^{-(x-\beta v)/2}(x - \beta v)^{-1/2} e^{-v/2} v^{-1/2}\, dv \\ &= k^2 e^{-x/2} \int_0^{x/2} e^{-(1-\beta)v/2}(x - \beta v)^{-1/2} v^{-1/2}\, dv. \end{aligned}$$

Putting $\beta v/x = t$, we have

$$f_\beta(x) = \beta \cdot k^2 e^{-x/2} \int_0^1 \exp\,(-(\beta^{-1} - 1)xt/2)(1 - t)^{-1/2} t^{-1/2}\, dt$$

which is strictly decreasing in x, since $(\beta^{-1} - 1)$ is positive.

Now by Theorem 3.1 and according to Definition 2.1, the procedure $S(\nu, c)$ is efficient.

Example 4.2. Let $\mathcal{C}$ be the class of Borel sets in $\mathbf{R}^2$ whose volume is atmost equal to d, a given positive number. For this class, the set C_0 which satisfies (2.1) is given by

$$E = \{x : x'\Sigma^{-1}x \leqslant r\}$$

where r is so chosen that the volume of E is d. The Minkovski metric corresponding to the set E is the metric $\rho(x, y)$ given by

$$\rho^2(x, y) = r^{-1}(x - y)'\Sigma^{-1}(x - y).$$

Now,

$$Pr\{Z \leqslant x\} = Pr\{U + V \leqslant rx\}$$
$$= H(rx, \Sigma).$$

It can be seen that the density of $U + V$ is strictly decreasing again by Lemma 4.1 and so by Theorem 3.1 the procedure $S(\nu, c)$ is efficient in the sense of Definition 2.1.

Example 4.3. We now take the class $\mathcal{C}$ to be the class of all ellipses of diameter $2d$. In this case the set C_0 is the ellipse

$$E = \{x : \lambda x'\Sigma^{-1}x \leqslant d\}$$

and the Minkovski metric corresponding to this set is

$$\rho^2(x; y) = d^{-1}(x - y)'\Sigma^{-1}(x - y).$$

We have

$$Pr\{Z \leqslant x\} = Pr\{U + V \leqslant 2d/\lambda\} = H(dx/\lambda; \Sigma).$$

The density of $H(\cdot; \Sigma)$ is strictly decreasing and so Theorem 3.1 is applicable and we obtain $S(\nu, c)$ as an efficient procedure.

Example 4.4. This is an example where Theorem 3.1 is not applicable. Let $\mathcal{C}$ be the class of all squares of side d where d is a given positive number. It has been shown by Dasgupta and Rattihalli (1983) that the set C_0 in this case is given by

$$C_0 = \{x : |x_1| + |x_2| \leqslant d\}.$$

The Minkovski metric in this case is

$$\rho(x, y) = d^{-1}(|x_1 - y_1| + |x_2 - y_2|).$$

It can be seen that the density of Z is not strictly decreasing. Theorem 3.1 is not applicable. We could not prove Theorem 3.1 without the assumption 2. It is not known whether the procedure $S(\nu, c)$ is efficient.

Remark 4.1. In higher dimensions, in most interesting cases the metric $\rho(\cdot)$ is as in Example 4.4, that is the density of Z is not strictly decreasing. In order to extend Theorem 3.1 to higher dimensions with wide enough applicability it is necessary to relax Assumption 2. We were not able to do this.

Remark 4.2. In Example 4.1 we took $\mathcal{C}$ to be the class of all *symmetric* convex sets of diameter $2d$ and the set C_0 in this case was the circle. When the condition of symmetry is removed, we are not able to identify the set C_0. We conjecture that it is still the circle and the conjecture is true when $\Sigma = I$, the identity matrix. The source of difficulty in proving the conjecture is that we could not evaluate the probability content of the circle in relation to other sets of constant width. Recently R. N.

Rattihalli (Personal Communication) has shown that the conjecture is true using Steiner symmetrization of convex sets and the fact that the bivariate normal density with diagonal Σ is decreasing in the square of each co-ordinate and so Steiner symmetrization does not increase the diameter of a convex set.

11.5 DEFINITION OF ASYMPTOTIC EFFICIENCY OF FIXED SIZE PROCEDURE

In the general set-up described in the beginning of this paper, we define the asymptotic efficiency of an asymptotically α-consistent fixed size region based on a statistic T_n as follows.

Let $\{T_n\}$ be a consistent asymptotically normal estimator of θ with limiting covariance matrix $\Sigma = \Sigma(P)$. Let $n_0(p, d) = n_0(\Sigma(P), d, \mathcal{C}(d))$ be the sample size required by the efficient procedure $S_d(\nu, c)$, if $\sqrt{n}(T_n - \theta)$ were distributed as $N(0, \Sigma(P))$ for every n. We may then have the following definition.

Definition 5.1. A family $\mathcal{S} = \{S_d\}$ of asymptotically α-consistent procedures based on T_n is said to be *asymptotically efficient* if

$$\lim_{d \downarrow 0} \frac{E_p N(d)}{n_0(p, d)} \leqslant 1$$

for all P in $\mathbb{P}$.

Williams and Sen (1974) considered (among other problems) the above problem with $\mathcal{C}(d)$ the class of all ellipses of diameter $2d$. For this choice of $\mathcal{C}(d)$ it is appropriate to consider procedures based on a best asymptotically normal estimator, if one is available.

Mukhopadhyaya (1981) considered the problem of constructing fixed size regions for the mean and variance of a normal distribution. He took $\mathcal{C}(d)$ to be the class of all sets of the type C, where C is a symmetric convex set of diameter $2d$ and Θ is the half-plane. His optimal procedure is not efficient according to our definition. The motivation for his definition of optimality is not clear.

Acknowledgements. The authors are thankful to the referee whose comments greatly improved the presentation and also to Dr. R. N. Rattihalli for making available to us a pre-print of his paper.

REFERENCES

1. Berger, J. O. (1980): *Statistical Decision Theory: Foundations, Concepts and Methods. Springer-Verlag*, New York.
2. Chow, Y. S. and Robbins, H. (1965): On the asymptotic theory of fixed width sequential confidence interval for the mean. *Ann. Math. Statist.*, **36**, 457–462.
3. Das Gupta, S. and Rattihalli, R. N. (1983): *Probability content of a rectangle under the normal distribution: Some inequalities*. Tech. Rep. No. **33/83**, Indian Statistical Institute.
4. Rockafellar, R. T. (1970): *Convex Analysis*, Princeton University Press.

5. Mukhopadhyay, N. (1981): Sequential approach to simultaneous estimation of the mean and variance, *Metrika*, **20,** 133–136.

6. Stein, C. and Wald, A. (1947): Sequential confidence interval for the mean of a normal distribution with known variance, *Ann. Math. Statist.*, **18**, 427–433.

7. Williams, G. W. and Sen, P. K. (1947): On bounded maximum width sequential confidence regions based on generalized *u*-statistics, *Jour. Multivariate Anal.*, **4,** 453–468.

8. Zacks, S. (1971): *The Theory of Statistical Inference*, John Wiley, New York.

9. Olmsted. J. M. H. (1970): *Advanced Calculus*, Eurasia Publishing House, New Delhi.

12

Iterated Logarithm Laws, Confidence Sequences and Sequential Tests

R. Vasudeva

University of Mysore, India

12.1 INTRODUCTION

Let $(X_n, n \geqslant 1)$ be a sequence of independent and identically distributed (i.i.d.) random variables (r.v.) defined on a common probability space and let $S_n = \sum_{j=1}^{n} X_j$, $n \geqslant 1$. Let $EX_1 = \mu$ when it exists. Define $T_n = S_n - n\mu$ when μ exists and $= S_n$ when μ does not exist. A sequence (b_n) is said to be an upper class sequence of (T_n) if $P(|T_n| \geqslant b_n \text{ i.o.}) = 0$. Iterated logarithm laws attempt to give precise upper class sequences. Here we present some iterated logarithm laws and discuss the boundary value problems associated with these laws. In section 12.2 we give some almost sure confidence sequences and sequential tests for the mean μ and variance σ^2, under certain assumptions on the r.v. X_1. We also present a criterion of selecting confidence sequences which satisfy some desirable boundary crossing conditions. In section 12.3, confidence sequence and sequential tests have been developed for μ and for the characteristic exponent α, when X_1 is a stable r.v. with $1 < \alpha < 2$. In the last section we give some applications of stable laws and make a brief note on the developments in the estimation procedures of the parameters of stable laws. This section may be treated as a supplement to section 12.3. We now present the version of the characteristic function of stable laws which will be repeatedly used in this paper.

A stable r.v. X_1 is described through its characteristic function (ch.f.) $\phi(t)$ as

$$\log \phi(t) = i\theta t - c|t|^{\alpha}\left(1 + \frac{i\beta t}{|t|} w(t, \alpha)\right),$$

where

$$w(t, \alpha) = \begin{cases} \tan(\pi\alpha/2) & \text{if } 0 < \alpha \leqslant 2, \alpha \neq 1 \\ \dfrac{2}{\pi} \log |t| & \text{if } \alpha = 1. \end{cases}$$

Here $c > 0, 0 < \alpha \leqslant 2, -1 \leqslant \beta \leqslant 1$ and $-\infty < \theta < \infty$ are the parameters of the distribution. α is called the characteristic exponent of the distribution. When $1 < \alpha \leqslant 2$, $EX_1 = \mu$ exists and further $\theta = \mu$. In particular, if $\alpha = 2$ then X_1 becomes a normal r.v. with mean μ and variance $2c$. When $0 < \alpha < 2$ and $\beta = 0$, X_1 becomes symmetric stable and when $0 < \alpha < 1$, $\beta = -1 (\beta = 1)$, X_1 becomes a positive (negative) valued r.v. When X_1 is stable with $0 < \alpha \leqslant 2$, $\alpha \neq 1$, we note that for each n, the distribution of $n^{-1/\alpha} T_n$ will be again stable with the same exponent α.

12.2. LAWS OF THE ITERATED LOGARITHM

We present here two iterated logarithm laws: (i) when $EX_1^2 < \infty$, and (ii) when X_1 is a stable r.v. with characteristic exponent α, $0 < \alpha < 2$, $\alpha \neq 1$. The case $\alpha = 1$ is omitted throughout the paper.

When $EX_1 = \mu$ and $\text{Var}(X_1) = \sigma^2$, $0 < \sigma^2 < \infty$, P. Hartman and A. Wintner (1941) established that

$$\limsup_{n \to \infty} \frac{|T_n|}{\sqrt{2n \log \log n}} = \sigma \text{ a.s.} \tag{1}$$

V. Strassen (1966) obtained the converse of (1). He established that if (1) holds for some μ and σ, $-\infty < \mu < \infty$, $0 < \sigma < \infty$, then $EX_1 = \mu$ and $\text{Var}(X_1) = \sigma^2$. From (1), we notice that for any $\gamma > 1$,

$$P(|T_n| > \gamma\sigma\sqrt{2n \log \log n} \text{ i.o.}) = 0 \tag{2}$$

i.e. $(\gamma\sigma\sqrt{2n \log \log n})$, $\gamma > 1$, form precise upper class sequences. We claim them to be precise because $(\gamma\sigma\sqrt{2n \log \log n})$ fails to be upper class sequences when $\gamma \leqslant 1$.

When X_1 is a non-normal stable r.v. with exponent $\alpha \neq 1$, [R. P. Pakshirajan and R. Vasudeva (1977)] we observe that

$$\limsup_{n \to \infty} |n^{-1/\alpha} S_n|^{1/(\log \log n)} = e^{1/\alpha} \text{ a.s.} \tag{3}$$

By (3) we have $(n^{(1+\delta)/\alpha})$, $\delta > 0$, as upper class sequences of (T_n) (when X_1 is a non-normal stable r.v. with $\alpha \neq 1$).

Associated with any upper class sequence (b_n), a r.v. of natural interest is the number of n for which $|T_n| \geqslant b_n$. Define $N(b_n) = \sum_{n=1}^{\infty} I(|T_n| \geqslant b_n)$, where $I(A)$ is the indicator function of a set A and observe that $N(b_n)$ gives the number of n's for which $|T_n| \geqslant b_n$. From the fact that $P(|T_n| \geqslant b_n \text{ i.o.}) = 0$, we note that $N(b_n)$ is a proper r.v. Under (2), J. Slivka (1969) established that $EN^t(b_n) = \infty$ for any $t > 0$. When X_1 is a stable r.v. with exponent α, $0 < \alpha < \infty$, $\alpha \neq 1$, and when $b_n = n^{1/\alpha}(\log n)^{\gamma/\alpha}$, $\gamma > 1$, by proceeding as in the lemma in J. Slivka (1969) one obtains that for any t with $0 < t < 1$,

$$EN^t(b_n) \geqslant \sum_{n=1}^{\infty} n^{(t-1)} P(|T_n| \geqslant b_n)$$

$$= \sum_{n=1}^{\infty} n^{(t-1)} P(|X_n| \geqslant (\log n)^{\gamma/\alpha}).$$

From the fact that X_1 is a stable r.v., we know that [see e.g. Feller (1969)] $(\log n)^\gamma P(|X_n| \geqslant (\log n)^{\gamma/\alpha})$ tends to some positive constant c_1 as $n \to \infty$.

Since $\sum_{n=1}^{\infty} n^{(t-1)} (\log n)^{-\gamma} = \infty$ we now conclude that

$$\sum_{n=1}^{\infty} n^{(t-1)} P(|X_n| \geqslant (\log n)^{\gamma/\alpha}) = \infty$$

and in turn that $EN^t(b_n) = \infty$, for any $t > 0$. T. L. Lai and K. K. Lan (1976c) have studied the finiteness of $EN^t(b_n)$ for various upper class sequences (b_n) and for various constants $t > 0$. We now present here Theorem 3 of T. L. Lai and K. K. Lan (1976c) and obtain its corollary which is needed in this paper.

Theorem. Let (X_n) be a sequence of i.i.d. r.v.'s with $EX_1 = 0$, $EX_1^2 = \sigma^2$, $E|X_1|^{2(r+1)}(1 + \log |X_1|)^{-(r+1)} < \infty$. Then

$$EN^r(\epsilon\sqrt{n \log n}) < \infty \quad \text{for all } \epsilon > (2r)^{1/7}\sigma.$$

Corollary. Let (X_n) be a sequence of i.i.d. r.v.'s with $EX_1 = 0$. If the moment generating function of X_1 exists and if $b_n = n^{1/2} (\log n)^{(1+\delta)/2}$ for some $\delta > 0$, then $EN^r(b_n) < \infty$ for all $r > 0$.

Proof. The corollary is immediate from the above theorem, since for any $\epsilon > 0$, however large it is,

$$\epsilon\sqrt{n \log n} = 0(\sqrt{n}\,(\log n)^{(1+\delta)/2}) \quad \text{for any } \delta > 0.$$

12.3. CONFIDENCE SEQUENCES AND SEQUENTIAL TESTS FOR MEAN AND VARIANCE

In this section we assume that the r.v. X_1 has finite variance σ^2. The technique of obtaining the confidence sequences and sequential tests of power 1 is due to D. A. Darling and H. Robbins (1967a). Since then we have a number of papers in this direction. To name a few, we have D. A. Darling and H. Robbins (1967b), (1967c), H. Robbins and D. Siegmund (1969), H. Robbins (1970), T. L. Lai (1976a), (1976b) and so on.

For any given confidence coefficient β, we now present the technique of obtaining confidence sequences which have probability β of simultaneously containing the true parameter, for every sample of size $n \geqslant m$, where m is an integer whose value depends on the confidence coefficient β and on the upper class sequence used to obtain the confidence sequences.

12.3.1 Confidence Sequences for μ When σ is Known

Let (a_n) be a sequence of positive numbers such that (σa_n) is an upper class sequence of (T_n). Then we have

$$P(|S_n - n\mu| \geqslant \sigma a_n \text{ i.o.}) = 0 \tag{4}$$

Let $P_m = P(|S_n - n\mu| \geqslant \sigma a_n$ for some $n \geqslant m)$. By (4) we know that $P_m \to 0$ as $m \to \infty$. Define

$$I_n = \left(\frac{S_n}{n} - \frac{\sigma a_n}{n}, \frac{S_n}{n} + \frac{\sigma a_n}{n}\right).$$

Then $P(\mu \in I_n$ for all $n \geqslant m) = 1 - P_m$. Hence $(I_n, n \geqslant m)$ is a confidence sequence for μ with confidence coefficient $1 - P_m$. Define $J_n = \bigcap_{k=m}^{n} I_k$. Then $(J_n, n \geqslant m)$ also forms a confidence sequence for μ with the same confidence coefficient $1 - P_m$. This sequence has the advantage that for each $n \geqslant m$, $J_n \subset I_n$ and $J_m \supset J_{m+1} \supset J_{m+2} \ldots$.

In order to obtain confidence sequences which meet a given confidence coefficient, we need obtain a bound for P_m. This bound depends on the sequence (a_n). Here we obtain such bounds when $a_n = \gamma\sqrt{2n \log \log n}$, $\gamma > 1$, $a_n = \sqrt{(n+1)(a^2 + \log(n+1))}$, $a > 0$ and $a_n = \sqrt{n}\,(\log n)^{(1+\delta)/2}$, $\delta > 0$, and compare the corresponding confidence sequences.

We observe that $a_n = \gamma\sigma\sqrt{2n \log \log n}$ gives a precise upper class sequence. We now give an upper bound for P_m for this sequence.

Define

$$f(n) = (2 \log \log n + 6 \log \log \log n + 6 \log 2)^{1/2} n^{-1/2} \exp\left(\frac{1}{2 \log \log n}\right), n \geqslant 9.$$

Under the assumption that the moment generating function $m(t)$ of $(X_1 - \mu) \mid \sigma$ exists and is $\leqslant \exp(t^2/2)$, D. A. Darling and H. Robbins (1967) established that for all $m \geqslant 9$,

$$P\left(\left|\frac{S_n}{n} - \mu\right| \geqslant \sigma f(n) \text{ for some } n \geqslant m\right) \leqslant \frac{1}{\log \log (m-1)}.$$

Given $a_n = \gamma\sqrt{2n \log \log n}$, one can find an integer N_1 such that $nf(n) \leqslant a_n$ for all $n \geqslant N_1$. Hence for all $m \geqslant N_1$ one gets

$$P(|S_n - n\mu| > \sigma a_n \text{ for some } n \geqslant m) \leqslant \frac{1}{\log \log (m-1)}.$$

Now given the confidence coefficient β, find the least integer $m_0 \geqslant N_1$ such that $[\log \log (m_0 - 1)]^{-1} \leqslant 1 - \beta$. Then $(I_n, n \geqslant m_0)$ forms a confidence sequence with confidence coefficient $1 - \beta$. In particular, if $(X_1 - \mu)/\sigma$ is $N(0, 1)$ then it may be noted that $m(t) = \exp(t^2/2)$. Hence these confidence sequences are valid when X_1 is a normal r.v.

When $a_n = \sqrt{(n+1)(a^2 + \log(n+1))}$, $a > 0$, under the same assumptions on the moment generating function $m(t)$, H. Robbins (1970) obtained that

$$P_1 = P(|S_n - n\mu| \geqslant \sigma a_n \text{ for some } n \geqslant 1) \leqslant \exp(-a^2/2).$$

Hence the confidence sequence I_n becomes

$$\left(\frac{S_n}{n}-\frac{\sigma\sqrt{(n+1)(a^2+\log(n+1))}}{n}, \frac{S_n}{n}+\frac{\sigma\sqrt{(n+1)(a^2+\log(n+1))}}{n}\right), n \geqslant 1.$$

Given a confidence coefficient β, 'a' can be chosen such that $P_1 \leqslant 1-\beta$.

When $a_n = \sqrt{n}\,(\log n)^{(1+\delta)/2}$, $\delta > 0$, under the weaker assumption that $Var\,(X_1)=\sigma^2$, is known. D. A. Darling and H. Robbins (1967b) established that

$$P_m = P(|S_n - n\mu| \geqslant \sigma a_n \text{ for some } n \geqslant m) \leqslant \frac{\sigma^2}{\delta\,(\log m)^\delta}\left(1+\frac{\delta}{\log m}\right).$$

Given a confidence coefficient β, one can find a m_0 such that

$$\frac{\sigma^2}{\delta\,(\log m_0)^\delta}\left(1+\frac{\delta}{\log m_0}\right) \leqslant 1-\beta.$$

Consequently a confidence sequence with confidence coefficient β is given by

$$\left(\frac{S_n}{n}-\frac{\sigma(\log n)^{(1+\delta)/2}}{\sqrt{n}}, \frac{S_n}{n}+\frac{\sigma(\log n^{(1+\delta)/2}}{\sqrt{n}}\right), n \geqslant m_0.$$

When $b_n = \gamma\sigma\sqrt{2n \log\log n}$, $\gamma > 1$, we observe that for each n, the length of the confidence interval is the shortest as compared to any other upper class sequence (b_n). This is in view of the fact that $\gamma\sigma\sqrt{2n\log\log n}$, $\gamma > 1$, are precise upper class sequences of (T_n). Eventhough, confidence sequences with intervals of shortest length are desirable, we now present a major drawback with such confidence sequences.

Given any upper class sequence (b_n), the associated confidence sequences are given by $\left(\frac{S_n - b_n}{n}, \frac{S_n + b_n}{n}\right)$, $n \geqslant m$ where m is an integer determined on the basis of the confidence coefficient β. For each such confidence sequence, let $N(b_n)$ denote the number of n's such that μ is not included in $\left(\frac{S_n - b_n}{n}, \frac{S_n + b_n}{n}\right)$. From the fact that $P(|T_n| \geqslant b_n \text{ i.o.}) = 0$, we note that $N(b_n)$ is a proper r.v. Further it can be represented as $N(b_n) = \sum_{n=m}^{\infty} I(|T_n| \geqslant b_n)$, where $I(A)$ is the indicator function of any set A. When $b_n = \gamma\sigma\sqrt{2n\log\log n}$, $\gamma > 1$, by Section 1, we observe that $EN^t(b_n) = \infty$ for any $t > 0$. Hence, eventhough the confidence intervals are of shortest length, we notice that the number of intervals

$$\left(\frac{S_n}{n}-\frac{\gamma\sigma\sqrt{2n\log\log n}}{n}, \frac{S_n}{n}+\frac{\gamma\sigma\sqrt{2n\log\log n}}{n}\right),$$

which do not include μ takes large values with high probability. This is a serious drawback of these confidence sequences. Similarly, when $b_n = \sigma\sqrt{(n+1)(a^2+\log(n+1))}$, by the theorem given in Section 12.2, we have $EN^t(b_n) < \infty$ if $t < 1/2$, $= \infty$ if $t \geqslant 1/2$. A criterian that is desirable of a good confidence sequence is that $EN^k(b_n) < \infty$ for as large a k

as possible. When $b_n = \sigma\sqrt{n}\,(\log n)^{(1+\delta)/2}$, $\delta > 0$, and when the moment generating function of X_1 exists then by the Corollary in Section 12.2, we notice that $EN^t(b_n) < \infty$ for all $t > 0$. Applying strong law of large numbers one may notice that this confidence sequence eventually shrinks to μ a.s. In view of the finiteness of $EN^t(b_n)$ for all $t > 0$, one may prefer the confidence sequence with $b_n = \sqrt{n}\,(\log n)^{(1+\delta)/2}$, $\delta > 0$, over the other two sequences obtained above. In particular, if the r.v. X_1 is $N(\mu, \sigma^2)$, then a bound for $EN(b_n)$ is presented below.

$$EN(b_n) = \sum_{n=m}^{\infty} P(|S_n - n\mu| > b_n) = 2 \sum_{n=m}^{\infty} \phi[-(\log n)^{(1+\delta)})/2]$$

where ϕ is the standard normal distribution function. But we know that $\phi[-((\log n)^{(1+\delta)})/2] \leqslant \exp[-((\log n)^{1+\delta})/2]$. Hence we have

$$EN(b_n) \leqslant \sum_{n=1}^{\infty} \exp[-(\log n)^{1+\delta}/2].$$

Unfortunately, there is a restriction on the minimal sample size needed, for using this confidence sequence.

We now compare the confidence sequence with

$$b_n = \sigma\sqrt{(n+1)(a^2 + \log(n+1))},\ a > 0,$$

with confidence intervals of fixed sample size. This choice of b_n is made because it gives confidence intervals for all $n \geqslant 1$. When the confidence coefficient $\beta = 0.95$ [H. Robbins (1970)], we have $a = \sqrt{6}$. The confidence sequence is given by

$$I_n = \left(\frac{S_n}{n} - \frac{\sigma\sqrt{(n+1)(6+\log(n+1))}}{n},\ \frac{S_n}{n} + \frac{\sigma\sqrt{(n+1)[6+\log(n+1)]}}{n}\right).$$

For a given sample size n, the best fixed sample size confidence interval with confidence coefficient 0.95 is $\left(\frac{S_n}{n} - \frac{1.96\sigma}{\sqrt{n}}, \frac{S_n}{n} + \frac{1.96\sigma}{\sqrt{n}}\right)$, which is included in I_n. By the iterated logarithm law in (1), one can observe that for any integer m

$$P\left(\frac{S_n}{n} - \frac{1.96\sigma}{\sqrt{n}} \leqslant \mu \leqslant \frac{S_n}{n} + \frac{1.96\sigma}{\sqrt{n}} \text{ for all } n \geqslant m\right) = 0.$$

This remains true if 1.96 is replaced by any other constant.

12.3.2. Confidence Sequence when both μ and σ are Unknown

D. A. Darling and H. Robbins (1967c) obtained the following confidence sequences for μ and σ under the assumption that the r.v. X_1 is $N(\mu, \sigma^2)$.

Let $\bar{X}_n = S_n/n$ and $V_n^2 = (n-1)^{-1} \sum_{i=1}^{n} (X_i - \bar{X}_n)^2$. Observe that distributionally $V_{n+1}^2 = \frac{\sigma^2}{n}(U_1 + U_2 + \ldots + U_n)$, where $U_1, U_2, \ldots, U_n$ are squares of independent $N(0, 1)$ r.v.'s. Set $Y_i = (1 - U_i)/\sqrt{2}$, $i = 1, 2, \ldots, n$.

Then the moment generating function $m(t)$ of Y_1 is given by

$$m(t) = \exp (t/\sqrt{2})(1 + \sqrt{2}\, t)^{-1/2} \leqslant \exp (t^2/2).$$

Hence for any upper class sequence (a_n) of $\sum_{i=1}^{n} Y_i$ we have (assuming that $a_n \uparrow$ and $a_n/n \downarrow$)

$$P_m = P\left(\left|\sum_{i=1}^{n} Y_i\right| \geqslant a_n \text{ for some } n \geqslant m\right)$$

$$= P\left(V_n^2 \leqslant \sigma^2\left(1 - \frac{\sqrt{2}\, a_{n-1}}{n-1}\right) \text{ for some } n \geqslant m + 1\right).$$

Let m_1 be an integer large enough such that $\sqrt{2}\, a_{n-1}/(n-1) < 1$ for all $n \geqslant m_1$. For $m \geqslant m_1$ define

$$W_n = \min_{m<i<n} V_i\left(1 - \frac{\sqrt{2}\, a_{i-1}}{i-1}\right)^{-1/2}$$

Then $P(\sigma > W_n$ for some $n \geqslant m + 1)$

$$\leqslant P\left(V_n^2 \leqslant \sigma^2\left(1 - \frac{\sqrt{2}\, a_{n-1}}{n-1}\right) \text{ for some } n \geqslant m + 1\right)$$

$$= P_m.$$

For various sequences (a_n) considered in (12.3.1), bounds for P_m can be obtained [as $m(t) \leqslant \exp (t^2/2)$]. Consequently for a given confidence coefficient β, a m_0 $(\geqslant m_1)$ can be chosen such that $(0, W_n)$, $n \geqslant m_0$ is a confidence sequence for σ with β as the confidence coefficient.

Further, D. A. Darling and H. Robbins (1967c) established that the sequence (J_n') defined by

$$J_n' = \bigcap_{k=m+1}^{n} \left(\frac{S_n - W_n a_{n-1}}{n}, \frac{S_n + W_n a_{n-1}}{n}\right), \; n \geqslant m + 1,$$

can be considered as a confidence sequence for μ when σ is unknown. Here m is an integer chosen such that (J_n'), $n \geqslant m + 1$, meets the given confidence coefficient. For all sequences (a_n) considered in (12.2.1), m can be chosen such that

$$P(\mu \in J_n' \text{ for all } n \geqslant m + 1) \geqslant \beta,$$

where β is the given confidence coefficient. This is possible in view of the fact that both the probabilities,

$$P(|S_n - n\mu| \geqslant \sigma a_n \text{ for some } n \geqslant m)$$

and

$$P\left(V_n^2 \leqslant \sigma^2\left(1 - \frac{\sqrt{2}\, a_{n-1}}{n-1}\right) \text{ for some } n \geqslant m + 1\right)$$

can be made as small as desired by choosing m large.

Remark 1. Under the same setup, if μ is known to be μ_0, say, then the confidence sequence for σ can be obtained on the following lines. Define

$V_{on}^2 = n^{-1} \sum_{i=1}^{n} (X_i - \mu_0)^2$ and observe that in distribution

$$V_{on}^2 = \frac{\sigma^2}{n} (U_1 + U_2 + \ldots + U_n),$$

where $U_1, U_2, \ldots, U_n$ are squares of independent $N(0, 1)$ r.v.'s. Let $W_{on} = \min_{m \leqslant i \leqslant n} V_{oi}(1 - \sqrt{2}\, a_i/i)$. Then proceeding as in the case of μ unknown, for a given confidence coefficient β, a m can be chosen such that $(0, W_{on})$, $n \geqslant m$, is a confidence sequence for σ with confidence coefficient β.

12.3.3. Power 1 Sequential Test for μ when σ is Known

Let (X_n) be a sequence of i.i.d. $N(\mu, 1)$ r.v.'s. To test the hypothesis $H_0 : \mu \leqslant \mu_0$ against the alternative $H_1 : \mu > \mu_0$, H. Robbins (1970) gave the criterion as: reject H_0 at the stage N if $N =$ first $n \geqslant m$ such that $S_n - n\mu_0 > a_n$ and continue taking the next observation otherwise. Here (a_n) is an upper class sequence under H_0 and m is an integer chosen such that the test procedure has the required size. For $\mu \leqslant \mu_0$, we have

$$\begin{aligned} P_\mu(\text{reject } H_0) &\leqslant P_{\mu_0}(\text{reject } H_0) \\ &= P_{\mu_0}(S_n - n\mu_0 \geqslant a_n \text{ for some } n \geqslant m). \end{aligned}$$

Hence for sequences (a_n) considered in (12.3.1), m can be chosen such that the tests have the required size. Also for $\mu > \mu_0$, we have

$$\begin{aligned} P_\mu(\text{accept } H_0) &= P_\mu(S_n - n\mu_0 < a_n \text{ for all } n \geqslant m) \\ &= P_\mu\left(\frac{S_n}{n} - \mu < \frac{a_n}{n} - (\mu - \mu_0) \text{ for all } n \geqslant m\right). \end{aligned}$$

By the strong law of large numbers, we know that $S_n/n \to \mu$ a.s. Hence, for all sequences (a_n) such that $a_n/n \to 0$ as $n \to \infty$ (which is true of all sequences (a_n) considered in 12.2.1)

$$P_\mu\left(\frac{S_n}{n} - \mu \leqslant \frac{a_n}{n} - (\mu - \mu_0) \text{ for all } n \geqslant m\right) = 0.$$

i.e., the power of the test is 1 at any $\mu > \mu_0$.

Remark 2. In order to test $H_0 : \mu = \mu_0$ against $H_1 : \mu > \mu_0$, observe that the test procedure given above can itself be adopted. Here again it turns out to be a power 1 test. In particular for $a_n = \gamma\sqrt{2n \log \log n}$, $\gamma > 1$. D. A. Darling and H. Robbins (1976b) have obtained a bound for the expected sample size needed for the test procedure to terminate, i.e.

$$E_\mu N \leqslant 2(\mu - \mu_0)^{-2} \log \log (\mu - \mu_0)^{-2}[1 + 0(1)].$$

Remark 3. Confidence sequences and sequential tests of power 1 have been obtained in literature under various setups. For instance, D. A. Darling and H. Robbins (1967c) obtain confidence sequences for the median of a distribution and the mean (variance) of a Poisson distribution. H. Robbins (1970) obtains confidence sequences for the parameter p of a Bernoulli distribution, T. L. Lai (1976a), (1976b) and (1977) obtains confidence sequences and sequential tests for parameters of a Gamma distribution, Uniform distribution and so on.

12.4 CONFIDENCE SEQUENCES AND SEQUENTIAL TESTS FOR PARAMETERS OF STABLE DISTRIBUTIONS

All the results in this section are author's own contribution to this paper. Here we assume that X_1 is a stable r.v. with its characteristic function (ch.f.) $\phi(t)$ given by $\log \phi(t) = i\mu t - |t|^\alpha$, μ real and $1 < \alpha < 2$.

Then we know that $EX_1 = \mu$. Also, we know that [see e.g. J. Mijnheer (1975)] $E|X_1 - \mu|^p < \infty\ (=\infty)$ whenever $p < \alpha\ (p \geqslant \alpha)$. Let $\mu_p = E|X_1 - \mu|^p$ for $p < \alpha$. We now give confidence sequences for μ when α is known and power 1 sequential tests for μ (α known) and α (μ known).

12.4.1 Confidence Sequences for μ when α is Known

By (3) we have for any given $\delta > 0$,

$$P(|S_n - n\mu| \geqslant n^{(1+\delta)/\alpha} \text{ i.o.}) = 0.$$

Hence $(n^{(1+\delta)/\alpha})$, $\delta > 0$, form upper class sequences. Define

$$P_m = P(|S_n - n\mu| \geqslant n^{(1+\delta)/\alpha} \text{ for some } n \geqslant m).$$

We know that $P_m \to 0$ as $m \to \infty$. Hence, given a confidence coefficient β if an integer m_0 can be chosen such that $P_{m_0} \leqslant 1 - \beta$ then a confidence sequence for μ with confidence coefficient β can be given as

$$\left(\frac{S_n - n^{(1+\delta)/\alpha}}{n}, \frac{S_n + n^{(1+\delta)/\alpha}}{n}\right), \ n \geqslant m_0.$$

We now obtain an upper bound for P_m in order to obtain m for any given confidence coefficient.

Let (Z_n) be a nonnegative sub-martingale with $EZ_n \leqslant d_n$, $n \geqslant 1$ and let (b_n) be any nondecreasing sequence of positive constants. Then Chow's generalization of Hajek-Renyi inequality [see e.g. H. Robbins (1970)] is given as

$$P(Z_n \geqslant b_n \text{ for some } n \geqslant m) \leqslant \frac{d_m}{b_m} + \sum_{n=m+1}^{\infty} \frac{d_n - d_{n-1}}{b_n} \tag{5}$$

Define $Z_n = |S_n - n\mu|^p$, $\max(1, \alpha/(1 + \delta)) < p < \alpha$. By B. Von Bahr and C. G. Esseen (1965), we have $EZ_n \leqslant 2n\mu_p$ which when applied in (5) gives

$$\begin{aligned} P_m &= P(|S_n - n\mu| \geqslant n^{(1+\delta)/\alpha} \text{ for some } n \geqslant m) \\ &= P(Z_n \geqslant n^{(1+\delta)p/\alpha} \text{ for some } n \geqslant m) \\ &\leqslant 2\mu_p m^{1-(1+\delta)p/\alpha} + 2\mu_p \sum_{n=m+1}^{\infty} m^{-(1+\delta)p/\alpha} \leqslant C\mu_p m^{1-(1+\delta)p/\alpha} \end{aligned}$$

where $C = 2(1 + \alpha/[(1 + \delta)p - \alpha])$.

For any given confidence coefficient β, m should be chosen such that $C\mu_p m^{1-(1+\delta)p/\alpha} \leqslant 1-\beta$.

12.4.2 Power 1 Sequential Test for μ when α is Known

To test the hypothesis $H_0 : \mu \leqslant \mu_0$ against $H_1 : \mu > \mu_0$ let the criterion be to reject H_0 at stage N if $N =$ first $n \geqslant m$ such that $S_n - n\mu_0 > n^{(1+\delta)/\alpha}$ and to continue taking one more observation otherwise. Here δ is a small, but fixed positive number, and m is an integer whose value depends on the size of the test.

When $\mu \leqslant \mu_0$, we have

$$P_\mu(S_n - n\mu_0 > n^{(1+\delta)/\alpha} \text{ for some } n \geqslant m)$$
$$\leqslant P_{\mu_0}(S_n - n\mu_0 > n^{(1+\delta)/\alpha} \text{ for some } n \geqslant m).$$

By (5) one gets

$$P_{\mu_0}(S_n - n\mu_0 > n^{(1+\delta)/\alpha} \text{ for some } n \geqslant m) \leqslant C\mu_p m^{1-(1+\delta)p/\alpha},$$

where p is some constant chosen such that $\max(1, \alpha/(1+\delta)) < p < \alpha$. Given the size γ_0 of the test procedure, m can be found such that $C\mu_p \leqslant \gamma_0 m^{(1+\delta)p/\alpha-1}$. From the fact that $S_n/n \to \mu$ a.s., by proceeding as in Section 12.3.3, the power of the test can be shown to be 1.

12.4.3 Power 1 Sequential Test for α when μ is Known

To test $H_0 : \alpha = \alpha_0$ against $H_1 : \alpha < \alpha_0$, let the criterion be to reject H_0 at Nth stage if $N =$ first $n \geqslant m$ such that $|S_n - n\mu| \geqslant n^{(1+\delta)/\alpha_0}$ and to continue taking one more observation otherwise. Here δ is a small but fixed positive constant and m is a positive integer which depends on the size of the test. Let γ_0 be the size of the test. Then m is chosen such that

$$P_{\alpha_0}(|S_n - n\mu| > n^{(1+\delta)/\alpha_0} \text{ for some } n \geqslant m) \leqslant \gamma_0.$$

Again by (5), we have for $\max(1, \alpha_0/(1+\delta)) < p < \alpha_0$,

$$P_{\alpha_0}(|S_n - n\mu| > n^{(1+\delta)/\alpha_0} \text{ for some } n \geqslant m) \leqslant C\mu_p m^{1-(1+\delta)p/\alpha_0}.$$

Hence, by choosing m such that $C\mu_p m^{1-(1+\delta)p/\alpha_0} \leqslant \gamma_0$, the test procedure is made to satisfy the size condition.

For $\alpha < \alpha_0$, we have

$$P_\alpha(\text{reject } H_0) = P_\alpha(|S_n - n\mu| \geqslant n^{(1+\delta)/\alpha_0} \text{ for some } n \geqslant m)$$
$$\geqslant P_\alpha(|S_n - n\mu| \geqslant n^{(1+\delta)\alpha_0} \text{ i.o.})$$
$$= P_\alpha(|n^{-1/\alpha}(S_n - n\mu)| \geqslant n^{(1+\delta)/\alpha_0 - 1/\alpha} \text{ i.o.}).$$

When $(1+\delta)/\alpha_0 - 1/\alpha < 0$ (i.e. $\alpha < \alpha_0/(1+\delta)$) by (3), one gets

$$P_\alpha(|n^{-1/\alpha}(S_n - n\mu)| > n^{(1+\delta)|\alpha_0 - 1/\alpha} \text{ i.o.}) = 1.$$

As a consequence, we have P_α (reject H_0) $= 1$ when $\alpha < \alpha_0/(1+\delta)$. Choosing δ small such that $\alpha_0/(1+\delta) \simeq \alpha_0$, the test procedure, can in practice be claimed to be a power 1 test.

Remark 4. The probability bounds obtained in this section involve $\mu_p = E|X_1 - \mu|^p$, $\max(1, \alpha/(1+\delta)) < p < \alpha$. Since $X_1 - \mu$ is symmetric stable with ch. f. $\phi(t) = \exp(-|t|^\alpha)$, theory says that once α is known, it should be possible to evaluate μ_p. But in view of the fact that the density of a stable distribution can only be expressed as an infinite series [see Lukacs (1970)], computation of μ_p becomes very hard. We hence present a simple upper bound for μ_p.

12.4.4 An Upper Bound for μ_p

Let f denote the probability density function of $X_1 - \mu$. Then we have $\mu_p = \int_{-\infty}^{\infty} |x|^p f(x)\, dx$, $p < \alpha$. By the symmetry of $X_1 - \mu$, one gets $\mu_p = 2\int_0^\infty x^p f(x)\, dx$. From the fact that $X_1 - \mu$ is unimodal with the mode at zero [see e.g. page 158, Lukacs (1970)] and from the representation for $f(x)$ [see, e.g. page 142, Lukacs (1970)] we have the mode given by $f(0) = \frac{1}{\pi}\sqrt{(1/\alpha + 1)}$. Also, expanding $f(x)$ upto second degree [see, page 148, Lukacs (1970)] one gets for $x > 1$,

$$f(x) = \frac{1}{\pi x}\left[\frac{\sqrt{(1+\alpha)}\sin(\pi\alpha/2)}{x^\alpha} + \mathrm{Re}\left(\frac{\theta I_2}{2x^{2\alpha}}\right)\right],$$

where $|\theta| \leqslant 1$ and $I_2 = \exp(-i\pi(\alpha - 1)(2\alpha + 1)/2\alpha)\sqrt{(1+2\alpha)}$.

Hence $f(x) \leqslant \dfrac{\sqrt{(1+\alpha)}}{\pi x^{1+\alpha}} + \dfrac{\sqrt{(1+2\alpha)}}{2\pi x^{1+2\alpha}}$, $x \geqslant 1$. Consequently,

$$\begin{aligned}\mu_p &= 2\int_0^1 x^p f(x)\, dx + 2\int_1^\infty x^p f(x)\, dx \\ &\leqslant \frac{2f(0)}{(p+1)} + \frac{2}{\pi}\int_1^\infty x^p\left(\frac{\sqrt{(1+\alpha)}}{x^{1+\alpha}} + \frac{\sqrt{(1+2\alpha)}}{2x^{1+2\alpha}}\right) dx \\ &= \frac{2\sqrt{(1+1/\alpha)}}{\pi(p+1)} + \frac{2}{\pi}\frac{\sqrt{(1+\alpha)}}{(\alpha+p)} + \frac{\sqrt{(1+2\alpha)}}{\pi(2\alpha - p)} = A_{\alpha,\,p}, \text{ say.}\end{aligned}$$

Remark 5. For a given confidence coefficient β in 12.4.1, m can be found such that $CA_{\alpha,p} \leqslant (1-\beta)m^{(1+\delta)p/\alpha - 1}$. Similarly for a given size of the test γ in 12.4.2 and 12.4.3 m can be found such that

$$CA_{\alpha,p} \leqslant \gamma m^{(1+\delta)p/\alpha - 1} (CA_{\alpha_0,\,p} \leqslant \gamma m^{(1+\delta)p/\alpha_0 - 1}).$$

Remark 6. In Section 12.3, we introduced that concept of choosing confidence sequences which in turn satisfy $EN^t(b_n) < \infty$ for t large. Under stable laws with $1 < \alpha < 2$, by theorem 5, T. L. Lai and K. K. Lan (1976) confidence sequences for μ satisfying the above condition turn out to be of the type

$$\left(\frac{S_n - n^{(t+\delta)/\alpha}}{n}, \frac{S_n + n^{(t+\delta)/\alpha}}{n}\right).$$

Observing that $S_n/n \to \mu$ a.s., one may notice that for $t \geqslant 2$, other-confidence sequences tend to $(-\infty, \infty)$ a.s., as $n \to \infty$, which is contrary to what is desired of confidence sequences (i.e. one would expect the confidence sequence to shrink to the point μ as the sample size increases). Hence the criteria of choosing confidence sequences in terms of finiteness of $EN^t(b_n)$ for t large, becomes meaningless here.

12.5 APPLICATIONS OF STABLE LAWS

Let $[X(t), t \geqslant 0]$ be a Brownian motion with $X(0) = 0$ and let T_a be the first time at which $X(t)$ crosses a point $a > 0$. Then T_a has been found to be a positive stable r.v. with exponent $\alpha = 1/2$ [see e.g. page 171, Feller (1966)]. Astronomer Holtsmark observed that the distribution associated with gravitational field of stars turns out to be symmetric stable with $\alpha = 3/2$. These distributions were obtained on the basis of some theoretical considerations found in the phenomenon. Fitting of stable distributions on the basis of a numerical data began with B. Mendelbrot (1963).

When observations of very large magnitude occur frequently, B. Mendelbrot (1963 observed that it is not proper to treat them as outliers, since excluding them takes away a significant part of the data. He claimed that often in such situations stable laws describe the random phenomenon appropriately. He found that the behaviour of stock market prices can be better described by stable laws. Fama (1963) observed that stable laws described log of stock price changes better than the normal law. He also observed that positive stable laws $[0 < \alpha < 1, \beta = -1]$ appear to be a good fit for the distribution of claims due to perils of burglary suffered by large business concerns. In all these situations simulation technique was used. Analytical developments of estimation and testing procedures under stable laws are still in the initial stages. This may be attributed mainly to the complicated infinite series representation of the density of stable laws. Further, W. H. DuMouchel (1973) established that if $X_1, X_2, \ldots X_n$ is a sample of n independent observations from a stable distribution with parameters (α, β, c, μ) and if $L(\alpha, \beta, C, \mu, X_1, X_2, \ldots, X_n)$ is the likelihood function then L has no maximum for $0 < \alpha \leqslant 2$, $-\infty < \mu < \infty$. To be precise, he proved that L reaches maximum if $(\alpha, \mu) \to (0, X_k)$, when X_k is any observation. J. S. Press (1972), A. S. Paulson, E. W. Holcomb and R. A. Leitch (1975) obtained estimates of parameters through the empirical ch. f.

Let $\phi_n(t) = \frac{1}{n} \sum_{j=1}^{n} \exp(itX_j)$ be the empirical ch. f., where $X_1, X_2, \ldots, X_n$ is a sample of n independent observations from a stable distribution with ch. f. $\phi(t)$ described in Section 12.2. For $0 < \alpha < 2$, $\alpha \neq 1$, observe from Section 12.2 that

$$\text{Re}\,[\log(\phi(t))] = -C|t|^{\alpha}$$

and

$$\text{Im}\,[\log(\phi(t))] = \mu t - C|t|^{\alpha-1}\beta t \tan(\pi\alpha/2).$$

J. S. Press (1972) considered $\phi(t)$ at two arbitrary points i_1 and t_2 and solved for α and C through the equations

$$\text{Re}\,[\log\,(\phi(t_i))] = -C|t_i|^\alpha,\ i = 1, 2.$$

In the solution for α and C he replaced ϕ by ϕ_n in order to obtain the estimated α_n and C_n of α and C. Similarly, solving for Im $[\log\,(\phi(t))] = \mu t - C|t|^{\alpha-1}\,\beta t \tan\,(\pi\alpha/2)$ at points $t = t_3$, t_4 and substituting $\phi_n(t_3)$, $\phi_n(t_4)$, α_n and C_n for $\phi(t_3)$, $\phi(t_4)$, α and C he obtained the estimates β_n and μ_n of β and μ. These estimates are not satisfactory as t_1, t_2, t_3 and t_4 are arbitary. However, in view of the fact that ϕ_n is consistent for ϕ, he claimed that all his estimates are consistent. He further established that α_n and $\log C_n$ are asymptotically normal with respective means α and $\log C$.

P. J. Brockwell and B. M. Brown (1981) considered estimation of α and C for positive stable populations. For any $u > 0$, they evaluated the inverse moment EX^{-u}. Given a sample $X_1, X_2, \ldots, X_n$ of n independent observations from the above population they defined $\bar{X}^{-u} = \frac{1}{n}\sum_{i=1}^{n} X_i^{-u}$. By equating EX^{-u} to $\bar{X}^{-u}$ at $u = t$ and $2t$, $t > 0$, they solved for α_n and C_n. They further obtained Cramer-Rao type bounds and gave iterative methods aimed at choosing t so that α_n and C_n have high efficiency.

Acknowledgement. The author thanks the referee for his valuable comments in the light of which the paper was revised.

REFERENCES

1. Brockwell, P. J. and Brown, B. M. (1981): High efficiency estimation for positive stable laws, *J. Amer. Stat. Assn.*, **Vol. 76**, pages 626–631.

2. Darling, D. A. and Robbins, H. (1967a): Iterated logarithm inequalities, *Proc. Nat. Acad. Sci.*, U.S.A., **Vol. 57**, pages 1188–1192.

3. ——— (1967b): Inequalities for the sequence of sample sums, *Proc, Nat. Acad. Sci.*, U.S.A., **Vol. 57**, pages 1577–1580.

4. ——— (1967c): Confidence sequences for mean, variance and median, *Proc. Nat. Acad. Sci.*, U.S.A., **Vol. 58**, pages 66–68.

5. W. H. Du Mouchel (1973): On the asymptotic normality of maximum likelihood estimator when sampling from a stable distribution, *Ann. Stat.*, **Vol. 1**, pages 948–957.

6. Fama, E. (1963): Mendelbrot and the Stable Paretian hypothesis, *J. Business*, **Vol. 36**, pages 420–429.

7. ——— (1971): Parametric estimates of symmetric stable distributions. *J. Amer. Stat. Assn.*, **Vol. 66**, pages 331–338.

8. Feller, W. (1966): *An introduction to probability theory and its applications*, **Vol. 2**, Wiley, NY.

9. Gnedenko, B. V. and Kolmogorov, A. N. (1954): *Limit distribution of sums of independent random variables*, Addison-Wesley, Cambridge, Mass.

10. Hartman, P. and Wintner, A. (1941): On the law of the iterated logarithm, *Amer. J. Math.*, **Vol. 63**, pages 169–176.

11. Lai, T. L. (1976a): On confidence sequences, *Ann. Stat.*, **Vol. 4**, pages 265–280.

12. ——— (1976b): Boundary crossing probabilities for sample sums and confidence sequences, *Ann. Prob.*, **Vol. 4**, pages 299–312.

13. Lai, T. L. and Lan, K. K. (1976c): On the last time and the number of boundary crossings related to the Strong law of large numbers and the law of the iterated logarithm, *Zeitschrift Wahrscheinlichkeitstheorie Ver. Geb.*, **Vol. 34**, pages 59–71.

14. Lai, T. L. (1977): Power-One test based on sample sums, *Ann. Stat.*, **Vol. 5**, pages 866–880.

15. Lukacs, E. (1970): *Characteristic functions*, Griffin, London.

16. Mendelbrot, B. (1963): The variation of certain speculative prices, *J. Business*, **Vol. 36**, pages 394–419.

17. Mijnheer, J. L. (1975): Sample path properties of stable processes, *Math. Centre Tracts*, No. **59**, Math. Centrum, Amsterdam.

18. R. P. Pakshirajan and R. Vasudeva (1977): A law of the iterated logarithm for stable summands, *Trans. Amer. Math. Soc.*, **Vol. 232**, pages 33–42.

19. Paulsan, A. S., Holcomb, E. W. and Leitch, R. A. (1975): The estimation of parameters of stable laws, *Biometrica*, **Vol. 62**, pages 163–169.

20. Press, J. S. (1972): Estimation of univariate and multivariate stable distributions, *J. Amer. Stat. Assn.*, **Vol. 67**, pages 842–848.

21. Robbins, H. and Siegmund, A. (1969): Probability distributions related to the law of the iterated logarithm, *Proc. Nat. Acad. Sci.*, U.S.A., **Vol. 62**, pages 11–13.

22. Robbins, H. (1970): Statistical methods related to the law of the iterated logarithm, *Ann. Math. Stat.*, **Vol. 41**, pages 1397–1409.

23. Slivka, J. (1969): On the law of the iterated logarithm, *Proc. Nat. Acad. Sci.*, U.S.A., **Vol. 63**, pages 289–291.

24. Strassen, V. (1966): A converse to the law of the iterated logarithm, *Zeitschrift Wahrscheinlichkeitstheorie Ver. Geb.*, **Vol. 4**, pages 265–268.

25. Von, Bahr B. and Esseen, C. G. (1965): Inequalities for the r-th absolute moment of a sum of random variables, $1 \leqslant r \leqslant 2$, *Ann. Math. Stat.*, **Vol. 36**, pages 299–303.

13

Characterizations Based on Arithmetic, Geometric and Harmonic Means of Failure Rates

Dilip Roy and S. P. Mukherjee

Burdwan University and Calcutta University

13.1. INTRODUCTION

Let X be a life time variable with density function $f(x)$, distribution function $F(x)$ and survival function $\bar{F}(x)$. Then the failure rate (FR) at the point x is defined by

$$r(x) = f(x)/\bar{F}(x). \tag{1}$$

It has been extensively treated in the literature. How to define various classes of distributions, viz. Increasing Failure Rate (IFR) distributions, Decreasing Failure Rate (DFR) distributions and Constant Failure Rate (CFR) distributions based on the monotonicity of $r(x)$ (see Barlow and Proschan) is fairly known. Some characterizing properties based on $r(x)$ are also available, e.g., $r(x)$ is constant if and only if X follows the exponential distribution.

In case $r(x)$ is not monotonic in x, then the two classes of distributions, viz., Increasing Failure Rate Average (IFRA) distributions and Decreasing Failure Rate Average (DFRA) distributions are defined on the basis of monotonicity of the Arithmetic Mean Failure Rate (AFR), which in turn is defined by

$$A(x) = \frac{1}{x}\int_0^x r(t)\,dt \tag{2}$$

No attempt has, however, been made to define and study classes of distributions based on monotonicity of Geometric Mean Failure Rate (GFR) defined as

$$G(x) = \exp\left(\frac{1}{x}\int_0^x \log r(t)\,dt\right) \tag{3}$$

and of Harmonic Mean Failure Rate (HFR) defined as

$$H(x) = \left(\frac{1}{x}\int_0^x \frac{1}{r(t)}\,dt\right)^{-1} \tag{4}$$

But before we try to define such classes of distributions we need to examine the importance of GFR and HFR in the characterization of some well-known life time distributions. In this paper attempt has been made to derive some characterizing properties of the exponential, Weibull and Pearsonian type XI (also known as the Pareto distribution of the second kind) distributions based on various interrelationships among the FR, AFR, GFR and HFR.

13.2 CHARACTERIZING PROPERTIES

Theorem 2.1. The following statements are equivalent:

(a) AFR = GFR

(b) AFR = HFR

(c) GFR = HFR, and

(d) X follows the exponential distribution.

Proof follows from the fact that (a) $\Rightarrow r(x) =$ constant which in turn implies (b), (c) and (d), and further (d) implies (a).

Theorem 2.2. If $r(x)$ is differentiable, then

(a) $A(x) = ar(x)$ for all x if and only if X follows the Weibull distribution with shape parameter $(1/a)$.

(b) $G(x) = br(x)$ for all x if and only if X follows the Weibull distribution with shape parameter log (e/b), and

(c) $H(x) = cr(x)$ $(c < 2)$ for all x, if and only if X follows the Weibull distribution with shape parameter $(2 - c)$.

Proof. (a) Let $A(x) = ar(x)$. Then from the definition (2) we get

$$\int_0^x r(t)\, dt = axr(x)$$

Differentiating both sides w.r.t. x, we have

$$r(x) = axr'(x) + ar(x)$$

or

$$r'(x)/r(x) = (1 - a)/ax$$

Integrating both sides w.r.t. x, we have

$$\log [r(x)] = ((1 - a)/a) \log x + \log k$$

where log k is a constant of integration,

or

$$r(x) = kx^{(1/a)-1}$$

or

$$r(x) = r(1)x^{(1/a)-1}$$

Thus X follows the Weibull distribution with shape parameter $(1/a)$.

To prove the converse is easy. For shape parameter $(1/a)$ for a Weibull distribution

$$r(x) = r(1)x^{(1/a)-1}$$

Hence

$$A(x) = ar(1)x^{(1/a)-1}$$
$$= ar(x).$$

Proofs for part (b) and (c) are similar.

Corollary 2.1. If $r(x)$ is differentiable then the following statements are equivalent:

(a) $A(x) = r(x)$ for all x,
(b) $G(x) = r(x)$ for all x,
(c) $H(x) = r(x)$ for all x, and
(d) X follows the exponential distribution.

Proof follows from Theorem 2.2 by noting that

$$a = b = c = 1.$$

Theorem 2.3. If $r(x)$ is monotonic then the following statements are equivalent:

(a) $A(x)$ is proportional to $G(x)$ for all x,
(b) $A(x)$ is proportional to $H(x)$ for all x,
(c) $G(x)$ is proportional to $H(x)$ for all x, and
(d) X follows the Weibull distribution, wherever these measures are defined.

Proof. Because $r(x)$ is monotonic, $r(x)/A(x)$ is either greater than 1 or less than 1 for all x. Similarly $r(x)/H(x)$ is either greater than 1 or less than 1 for all x. Now consider the statement (a) of Theorem 2.3.

Let $A(x) = cG(x)$, where c is less than 1. The case $c = 1$ has already been considered in Theorem 2.1. Then

$$\int_0^x \log r(t)\, dt = x \log c - x \log x + x \log \int_0^x r(t)\, dt.$$

Differentiating both sides w.r.t. x, we have

$$\log r(x) = \log c - 1 - \log x + \log \int_0^x r(t)\, dt + x\left(\int_0^x r(t)\, dt\right)^{-1} r(x)$$

or

$$\log r(x) = \log (c/e) + \log A(x) + r(x)/A(x)$$

or

$$\log [r(x)/A(x)] = \log (c/e) + r(x)/A(x)$$

Writing $z = \log [r(x)/A(x)]$, we get

$$e^z - z = \log (e/c).$$

But $(e^z - z)$ is an increasing function for $z > 0$ and a decreasing function for $z < 0$ with a global minimum equal to 1 at $z = 0$. Now, because $\log (e/c)$ is greater than 1, we have either $z = z_1$ (> 0) for all values of x or $z = z_2$ (< 0) for all values of x because z is either greater than zero or less than zero for all values of x. Then an application of Theorem 2.2 implies that X follows the Weibull distribution. Thus statement (a) implies statement (d). Next consider statement (b). Let $H(x) = d \cdot A(x)$ where d is a constant less than 1. The case $d = 1$ has been taken into consideration in Theorem 2.1. Then

$$d \int_0^x r(t)\, dt = x^2 \left(\int_0^x \frac{1}{r(t)}\, dt \right)^{-1}$$

Differentiating both sides w.r.t. x, we have

$$d \cdot r(x) = 2x \left(\int_0^x \frac{1}{r(t)}\, dt \right)^{-1} + x^2 \left(\int_0^x \frac{1}{r(t)}\, dt \right)^{-2} \left(-\frac{1}{r(x)} \right)$$

or

$$d \cdot r(x) = 2H(x) - [H(x)]^2/r(x)$$

or

$$[H(x)/r(x)]^2 - 2[H(x)/r(x)] + d = 0$$

Writing $z = H(x)/r(x)$, we get

$$z^2 - 2z + d = 0,\ d < 1.$$

Then either $z = 1 + \sqrt{1-d}$ for all values of x or $z = 1 - \sqrt{1-d}$ for all values of x because of the choice of z and the fact that z is either > 1 or < 1. This implies that either

$$H(x) = (1 + \sqrt{1-d})r(x) \quad \text{for all } x$$

or

$$H(x) = (1 - \sqrt{1-d})r(x) \quad \text{for all } x.$$

Now an application of Theorem 2.2 proves that X follows the Weibull distribution. Thus statement (b) implies statement (d).

Proof for implication of statement (d) from statement (c) follows the same line as that adopted for implication of statement (d) from statement (c).

It may further be observed that statement (d) implies that for scale parameter λ and shape parameter α

$$r(t) = \lambda \alpha t^{\alpha-1}$$

and hence

$$A(x) = \lambda x^{\alpha-1}$$
$$G(x) = x^{\alpha-1} e^{\lambda\alpha-\alpha+1}$$
$$H(x) = \lambda\alpha(2-\alpha)x^{\alpha-1} \quad \text{(defined for } \alpha < 2).$$

Thus whenever such measures are defined statement (d) implies statements (a), (b) and (c).

This completes the proof of the theorem.

Theorem 2.4. If $r(x)$ is defined over $(0, \infty)$ and the second derivative of $r(x)$ exists, then $H(2x) = r(x)$ for all x, if and only if x follows Pearsonian type XI distribution.

Proof. Let $f(x) = 1/r(x)$. Then from the given relation, we have

$$2xf(x) = \int_0^{2x} f(t)\, dt.$$

Differentiating both sides w.r.t. x, we have

$$2f(x) + 2xf'(x) = 2f(2x)$$

or

$$f(2x) - f(x) = xf'(x) \quad \text{for all } x.$$

Now an application of mean value theorem implies that

$$f''(x) = 0 \quad \text{for all } x$$

or

$$f(x) = a + bx$$

or

$$r(x) = (a + bx)^{-1}$$

Regarding the choice of the constants a and b we conclude that (i) $a > 0$ from the fact $r(x) > 0$ and is finite at $x = 0$, and (ii) $b > 0$ because otherwise there could be a choice of $x = -(a/b)$ for which $r(x)$ would have been undefined.

Thus $r(x)$ represents the hazard rate for a Pearsonian type XI distribution. To prove the converse, we observe that for a Pearsonian type XI distribution with p.d.f.

$$p(x) = (q/c)(1 + x/c)^{-q-1}, \quad q > 0,\ c > 0$$

hazard rate

$$r(x) = (q/c)(1 + x/c)^{-1}$$

so that

$$H(x) = (q/c)(1 + x/2c)^{-1}$$

and hence

$$H(2x) = r(x) \text{ for all } x.$$

This completes the proof of the theorem.

13.3. CLASSES OF DISTRIBUTIONS

In view of the characterizing properties of the measures AFR, GFR and HFR discussed under Section 13.2, it may be worth defining classes of life time distributions on the line of IFRA (or DFRA) class of distributions.

Definition 1. A life time distribution is said to belong to the class of Increasing Geometric mean Failure Rate (IGFR) distributions if the GFR measured through $G(x)$ is an increasing function in x.

Definition 2. A life time distribution is said to belong to the class of Decreasing Geometric mean Failure Rate (DGFR) distributions if the GFR measured through $G(x)$ is a decreasing function in x.

Definition 3. A life time distribution is said to belong to the class of Increasing Harmonic mean Failure Rate (IHFR) distributions if the HFR measured through $H(x)$ is an increasing function in x.

Definition 4. A life time distribution is said to belong to the class of Decreasing Harmonic mean Failure Rate (DHFR) distribution if the HFR measured through $H(x)$ is a decreasing function in x.

Theorem 3.1. IFR class of distributions is a subset of each of the IFRA, IGFR and IHFR classes of distributions and DFR class of distributions is a subset of each of the DFRA, DGFR and DHFR classes of distributions.

Proof. If $r(x)$ is increasing (or decreasing) in x so are $A(x)$, $G(x)$ and $H(x)$ and hence follows the theorem.

Theorem 3.2. IFRA is a subset of IGFR which in turn is a subset of IHFR and DHFR is a subset of DGFR which in turn is a subset of DFRA.

Proof. From the definition of $A(x)$, $G(x)$ and $H(x)$, we get

$$A'(x) = -\frac{1}{x^2}\int_0^x r(t)\,dt + \frac{r(x)}{x}$$

or

$$A'(x) = [r(x) - A(x)]/x$$

Hence, $A(x)$ is an increasing (or decreasing) function in x implies

$$r(x) > (\text{or} <) A(x)$$

Similarly,

$$G'(x) = \left\{\exp\left(\frac{1}{x}\int_0^x \log r(t)\,dt\right)\right\}\left\{-\frac{1}{x^2}\int_0^x \log r(t)\,dt + \frac{1}{x}\log r(x)\right\}$$

or

$$G'(x) = \left\{\exp\left(\frac{1}{x}\int_0^x \log r(t)\,dt\right)\right\}\{\log r(x) - \log G(x)\}(1/x).$$

Hence $G(x)$ is an increasing (or decreasing) function in x implies

$$\log r(x) > (\text{or} <) \log G(x)$$

or

$$r(x) > (\text{or} <) G(x).$$

Also,

$$H'(x) = \left(\int_0^x \frac{1}{r(t)}\,dt\right)^{-1} - x\left(\int_0^x \frac{1}{r(t)}\,dt\right)^{-2}\frac{1}{r(x)}$$

or

$$H'(x) = \frac{H(x)}{x} - \frac{H^2(x)}{xr(x)}$$

or

$$H'(x) = \frac{H(x)}{x}\left(1 - \frac{H(x)}{r(x)}\right).$$

Hence $H(x)$ is an increasing (or decreasing) function in x implies

$$r(x) > (\text{or} <) H(x).$$

Now in case of IFRA, we have

$$r(x) > A(x)$$

and hence from the property that $A(x) > G(x)$

$$r(x) > G(x).$$

Thus IFRA $\subset$ IGFR. Similarly we can prove that IGFR $\subset$ IHFR and hence

$$\text{IFRA} \subset \text{IGFR} \subset \text{IHFR}.$$

Proceeding in the same way, we get

$$\text{DFRA} \supset \text{DGFR} \supset \text{DHFR}$$

This completes the proof of the theorem.

The above theorems thus establish the importance of these newly defined classes of life time distributions, specially the IHFR and IGFR classes.

We now consider a few more results regarding the IHFR and IGFR classes.

Theorem 3.3. (i) $\{\exp\}^{\text{CS, LD}} \subset$ IGFR and
(ii) $\{\exp\}^{\text{CS, LD}} \subset$ IHFR

Proof. The result follows from a combination of theorem 3.2 and a part of theorem 2.13 (p. 88) of Barlow and Proschan.

Let us write $a(x) = \exp[A(x)]$ and $h(x) = \exp[1/H(x)]$. Then from the fact that IFRA implies NBUE, one can observe that if X is IFRA, then

$$\int_t^\infty (x-t)\, d\{a(x)^{-x}\} \geqslant a(t)^{-t} \int_0^\infty x\, d\{a(x)^{-x}\}.$$

A similar result for IHFR class is given in the following theorem.

Theorem 3.4. If X is IHFR, then

$$\int_t^\infty (x-t)\, d\{h(x)^{-x}\} \leqslant h(t)^{-t} \int_0^\infty x\, d\{h(x)^{-x}\},$$

and the inequality is sharp.

Proof. Let us construct a random variable V such that the corresponding failure rate $r_V(x)$ is given by

$$r_V(x) = 1/r_X(x)$$

Then it is easy to express the cdf of V as

$$\bar{F}_V(x) = \exp\left\{-\int_0^x \frac{1}{r_X(u)}\, du\right\}$$
$$= \exp\{-x/H(x)\}$$

Further, X is IHFR implies that V is DFRA, which in turn implies that V is NWUE. Now an application of NWUE property implies that

$$\int_t^\infty \frac{(x-t)}{\bar{F}_V(t)}\, d\,\bar{F}_V(x) \leqslant \int_0^\infty x\, d\bar{F}_V(x)$$

or

$$\int_t^\infty \frac{(x-t)\, d\{\exp(-x/H(x))\}}{\exp\{-t/H(t)\}} \leqslant \int_0^\infty x\, d\{\exp(-x/H(x))\}$$

or

$$\int_t^\infty (x-t)\, d\{h(x)^{-x}\} \leqslant h(t)^{-t} \int_0^\infty x\, d\{h(x)^{-x}\}.$$

That the inequality is sharp follows from the fact that for the exponential distribution equality holds.

Similarly one can derive a corresponding result for IGFR class.

Acknowledgement. The authors thank the referee for suggesting the part of our work contained in theorems 3.3 and 3.4.

REFERENCES

1. Barlow, R. E. and Proschan, F. (1975): Statistical Theory of Reliability and Life Testing, Probability Models, Holt, Rinehart and Winston Inc.

14

Single-Arrival Bulk-Service Queues with One Server

M. L. Chaudhry
Royal Military College, Canada

J. Medhi
Formerly of Gauhati University, India

and

J. G. C. Templeton
University of Toronto, Canada

14.1 INTRODUCTION

Bailey (1954) was the first to introduce the concept of bulk service in queues. During the last 30 years there has been rapid development in this area, and bulk service systems have been put forward as models for many real life problems arising in transportation engineering and telecommunications. Various bulk service rules have been considered; the rules frequently discussed in the literature include the following:

(a) The server has a fixed maximum capacity, say b. If, immediately after completion of a service, the server finds more than b units waiting, he takes a batch of b for service while others wait in the queue and if he finds not more than b units, he takes all the available units in the batch for service. Bailey (1954) and Downton (1956) consider the rule that the server never relaxes and Bloemena (1960), Jaiswal (1960a, b) and others consider the rule that the server will not start service until there is at least one unit available for service.

(b) The server may take in a batch a fixed number k; he waits till the queue builds up to k and if there are more than k, he takes a batch of k while others wait. Fabens (1961), Takàcs (1962) and others consider this situation.

(c) The server may take in a batch a minimum of, say, a units and a maximum of, say, b units. If on completion of service of a batch, the server finds less than a units, he waits until there are a units, whereupon he takes a batch of a units for service. If he finds k units ($a \leqslant k \leqslant b$), he takes all the k units in a batch for service and if he finds more than b, he takes in a batch b units for service while others wait. Following

Neuts (1967), several researchers, such as Medhi (1979), Borthakur (1971, 1975), Easton and Chaudhry (1982), Kambo and Chaudhry (1982, 1983, 1984), Sim and Templeton (1982), and Chaudhry and Templeton (1983) have considered this rule. This rule will be called a general bulk service rule, which also covers the rules under (a) and (b) above as special cases.

(d) The capacity of the server may vary from batch to batch and may be given by a random variable Y. Bhat (1964), Teghem *et al.* (1969), Cohen (1969) and others considered this rule.

A single arrival bulk service model (with general arrival and service time distributions) under rule (c) is denoted by $GI/G(a, b)/1$ and that under rule (d) by $GI/G^Y/1$.

It is presumed that no batch will be accessible in the sense that once the service has started for a batch with less than the maximum capacity, no further unit will be admitted in the batch even if some arrival occurs during the service time. Newell (1960, 1965) considers a bulk service queue with accessible batches as a model of a traffic light queue.

Several aspects of bulk queues have been discussed in detail in the books by Bhat (1968), Teghem *et al.* (1969), Neuts (1981), Chaudhry and Templeton (1983), Medhi (1984a). Here we shall confine our attention to some of the other aspects of bulk service queues which have not yet appeared in any book.

14.2 MARKOVIAN MODELS

Such models which are analytically tractable have been investigated by several researchers using Erlangian methods. Borthakur (1971, 1975) studies the transient and steady state behaviour of the queue length (i.e., number of customers in the queue) for the system $M/M(a, b)/1$ while Medhi (1975, 1979) investigates the waiting time distribution in the steady state, and derives expressions for the moments of the waiting time. Medhi also verifies that Little's result holds for this system. Nadarajan and Subramanian (1984) consider the $M/M(a, b)/1$ system with removable server such that the server, when he finds less than a customers waiting on completion of a service, takes a vacation whose duration is given by an exponential random variable. The steady-state system state probability distribution is obtained in a modified matrix-geometric form. Finch (1962) considers the system $M/M(k, k)/1$ with fixed batch size k and limited waiting space. He investigates the transient and steady state distributions of the number in the system immediately after the completion of the nth service. Cosmetatos (1983) examines the system $M/M(1, b)/1$ and discusses the effect of server-sharing as a factor for increasing productivity. Easton and Chaudhry (1982) consider the system $E_k/M(a, b)/1$ in steady state and discuss its numerical analysis; Kambo and Chaudhry (1982) investigate its busy-period distribution. These Markovian systems have been investigated in some detail, explicit expressions have been found and numerical analyses have been suggested for important cases. We shall consider the more general case hereafter, from which corresponding results for the Markovian cases can be obtained as particular cases.

14.3 MODELS WITH POISSON INPUT AND GENERAL SERVICE TIME

14.3.1 Fixed Batch Size

The transient behaviour of the system $M/G(k, k)/1$ with fixed batch size has been studied by Takács (1962), Fabens (1961), Foster and his collaborators (1961, 1964). Fabens (1961) applies the theory of semi-Markov processes to find the steady-state distribution of the number in the system. This is done in two stages: first by determining the limiting distribution of departure epoch system size and then by filling up the remaining steps by renewal theory and certain results of semi-Markov process theory [see also Fabens and Perera (1963) for some corrections].

14.3.2.1 General bulk service rule, system size study

Neuts (1967) considers bulk queues with general bulk-service rule (as explained in 1(a)) and obtains a number of interesting transient and steady state results, both in discrete and in continuous time, for the system $M/G(a, b)/1$. In this system the service time for a batch of k $(a \leqslant k \leqslant b)$ customers has the d.f. $B_k(t)$ with L.T. $b_k^*(s)$ and mean α_k. Let X_n be the length of time between the departures of the $(n-1)$th and nth batches, $T_n = X_1 + X_2 + \ldots + X_n$, N_n^+ the number in the system immediately after departure of the nth service batch. The sequence $\{N_n^+, n \geqslant 0\}$ is a semi-Markov sequence. He obtains the g.f. of the L.T. of

$$Q_{ij}^{(n)}(x) = Pr\{N_n^+ = j,\ T_n \leqslant x \mid N_0^+ = i\}.$$

Let $p_{ij}^{(n)}$ be the n-step transition probability of the imbedded Markov chain $\{N_n^+, n \geqslant 0\}$. The Markov chain is positive recurrent *iff* $\lambda\alpha_b/b < 1$.

Let $p_j^+ = \lim_{n\to\infty} p_{ij}^{(n)}$ be the steady-state post-departure state probability. It is shown that the p.g.f. $P^+(z)$ is given by

$$\begin{aligned}
P^+(z) &= \sum_{j=0}^{\infty} p_j^+ z^j \\
&= \sum_{r=0}^{a-1} [p_r^+ z^b b_a^*(\lambda - \lambda z) - z^r b_r^*(\lambda - \lambda z) \\
&\quad + \sum_{r=a}^{b-1} p_r^+ z^b b_r^*(\lambda - \lambda z) - z^r b_b^*(\lambda - \lambda z)][z^b - b_b^*(\lambda - \lambda z))]^{-1}
\end{aligned}$$

which corresponds to the Pollaczek-Khinchine formula for $M/G/1$. The above formula for $P^+(z)$ can also be obtained by extending the argument of Kendall (1951) for obtaining the steady-state departure epoch distribution for $M/G/1$. In the particular case when the service time is independent of batch size, i.e., when $B_k(t) = B(t)$ for all k, the above formula reduces to

$$P^+(z) = \sum_{r=0}^{b-1} p_r^+(z^b - z^r)/[z^b/b^*(\lambda - \lambda z) - 1].$$

Neuts further investigates the continuous time distribution, both transient and steady state, of the system size $N(t)$ at time t. Denote

$$P_{ij}(t) = Pr\{N(t) = j \mid N(0) = i\}$$

and its L.T. by $p_{ij}^*(s)$. The distribution of $\{p_{ij}(t)\}$ is obtained from the renewal functions for the imbedded semi-Markov process $\{N_n^+, x_n, n \geqslant 0\}$. An explicit expression for the generating function $\sum_{j=0}^{\infty} P_{ij}^*(s)z^j$ is obtained.

The steady state exists *iff* $\lambda\alpha_b < b$. Explicit expressions for the steady state probability $p_j = \lim_{t\to\infty} p_{ij}(t)$ are given for the three cases $0 \leqslant j \leqslant a-1$, $a \leqslant j \leqslant b-1$, $j \geqslant b$. The expressions involve μ_j^*, the mean recurrence time of the state j in the imbedded semi-Markov process. It is observed that there does not exist any simple relation between p_j^* and p_j. It may be noted that $p_j^+ = p_j$ holds for $GI/G/1$ and that the relation between p_j^+ and p_j for $G/G(k, k)/1$ [as shown by Foster (1964)] is given by

$$p_j = (1/k) \sum_{i=0} p_i^+, \; j < k$$

$$= (1/k) \sum_{i=0}^{k-1} p_{j-i}^+, \; j \geqslant k$$

Neuts considers the busy period process

$$G_{ij}(x) = Pr\{I_n = j, y_n \leqslant x \mid I_{n-1} = i\}, \; n = 1, 2, \ldots$$

where Y_n is the length of the nth busy period, and I_n the number left behind $(0, 1, \ldots, a-1)$ at the end of nth busy period. Denote the L.T. of G_{ij} (which is independent of i) by $E_j(1, s), j = 0, 1, \ldots, a-1$. It is shown that E_j can be obtained as the solution of a system of linear equations. A method for obtaining the moments of the busy period is also indicated. A straightforward extension of the approach of Neuts has been used by Nadarajan and Sankaranarayanan (1980) to investigate $E_k/G(a, b)/1$. In this connection, see also Holman *et al.* (1981) who attempt to give numerical analysis of the simpler system $M/G(a, b)/1$.

14.3.2.2. General bulk service: Queue size study

Borthakur and Medhi (1974) investigate the system further. Apart from basic differences in the approach, their investigation relates to distribution of the number of customers in the queue (excluding those in a service batch, if the server is busy). They consider the more general $M^X/G(a, b)/1$ system using the well-known supplementary variable technique; reference to this and many other related queues are now available in many books; see, e.g., Chaudhry and Templeton (1983).

The busy period. Using the supplementary variable technique and considering the relevant equations, Borthakur (1975) obtains the distribution of the busy period T for the system $M^X/G(a, b)/1$ a special case of which, viz. $M/G(a, b)/1$, is discussed below. Let T be the interval from the instant of commencement of service of a batch leaving r in the queue to the first subsequent instant when the server becomes free.

Let

$$f_r(t)\,dt = Pr\{t \leqslant T < t + dt, I = r\},$$

where I is the number left behind the termination of the previous busy period $(0 \leqslant r \leqslant a - 1)$. The expression for the L.T. of the p.g.f. of the queue length during the busy period is given by

$$U^*(s, z) = \frac{1 - b^*(s + \lambda - \lambda z)}{s + \lambda - \lambda z} \cdot \frac{A_1(s, z)}{z^b - b^*(s + \lambda - \lambda z)},$$

where,

$$A_1(s, z) = z^{b+q} + \sum_{r=a}^{b-1} (z^b - z^r) \int_0^\infty U_r^*(x, s)\eta(x)\,dx - \sum_{r=0}^{a-1} z^r \int_0^\infty U_r^*(x, s)\eta(x)\,dx.$$

Applying Rouché's theorem one gets a set of linear equations from which $f_r^*(s) = \int_0^\infty U_r^*(x, s)\eta(x)\,dx$, $r = 0, 1, \ldots, b - 1$ can be obtained.

Borthakur considers the special case $G \equiv M$. With some correction, his results, for $q = 0$, lead to the L.T. of the p.d.f. of T as given by

$$f^*(s) = \sum_{r=0}^{a-1} f_r^*(s) = \frac{\mu[\{\alpha(s)\}^a - 1]}{(s + \mu)[\{\alpha(s)\}^a - \mu]},$$

where $f_r^*(s)$ is the L.T. of $f_r(t)$ and $\alpha(s)$ is the unique root of modulus greater than unity of the equation

$$z^{b+1} - (\lambda + \mu + s)z + \mu = 0.$$

Employing the method of Luchak (1958), one can get the inversion $f(t)$ of $f^*(s)$.

The above result has also been deduced as a particular case of their study of $E_k/M(a, b)/1$ by Kambo and Chaudhry (1982) (who incidentally point out an error in Borthakur's results).

In a subsequent paper Borthakur (1981) investigates the joint distribution of the length of the busy period, the queue length and the number of batches served during the busy period. He considers that the state of the system at epoch t involves one additional variable m, the number of batches served in the interval $(0, t)$, within a busy period; in place of $V_r(t)$ and $U_n(x, t)$, he considers $V_{r,m}(t)$ and $U_{n,m}(x, t)$ respectively. The treatment is then essentially the same. Several results obtained earlier can be deduced as particular cases.

14.3.2.3. System with variable service capacity

Let Y_n be the variable capacity for service ending at the $(n+1)$th service completion epoch $(n = 0, 1, \ldots)$. Assume that the r.v.'s $Y_1, Y_2, \ldots$ are i.i.d. with p.g.f. $g(z)$, and m is the largest positive integer such that $P(Y_n = m) > 0$. If at the n-th service completion the queue length is less than the current service capacity Y_n, then the server may have a number of alternatives: the server may take the available units for service, or wait for a minimum number or wait till the full current capacity is attained. The imbedded Markov chain analysis for the $M/G^Y/1$ system will be the same in all these cases. Such systems have been investigated by Bhat (1964, 1968), Teghem *et al.* (1969), Prabhu (1965). In this connection, see also Jaiswal (1961). The case of limited waiting space has been investigated by Singh (1971); he considers the system $M/G^Y/1, K$ having waiting space for K customers in addition to the batch being served; the capacity of the server is $b - Y_n$, where b is a fixed number and Y_n is a random variable. Singh obtains the generating function $Q(z)$ for the stationary distribution v_j of the departure epoch queue length j for $j = 0, 1, \ldots, k-1$ (Singh (1971), Bagchi and Templeton (1973a), Chaudhry and Templeton (1983)).

Following the approach of Cohen (1969), Bagchi and Templeton (1973b) investigate the transient and stationary behaviour of such a system with limited waiting space and having compound Poisson input. They consider the situation that the server never relaxes and that he takes in a batch for service the whole queue length or the current capacity of the server whichever is less. Let Q_n denote the number of customers in the system immediately after the completion of the n-th service batch; then the p.g.f. $Q(z)$ is given by

$$Q(z) = \frac{1-z}{g(1/z)b^*(\lambda - \lambda z)} \frac{f(z)b^*(\lambda - \lambda z)}{z^m} + \frac{h(z)z^K}{b_2^*(\lambda - \lambda z)}$$

where the L.T. of the service time $b^*(z)$ is a rational function of the form $b^*(z) = b_1^*(z)/b_2^*(z)$, the degree of b_1 being $n-1$ or less, and that of b_2 being n, and $f(z)$ and $h(z)$ are polynomials of degree $m-1$ and $n-1$ respectively, such that

$$\frac{f(z)b^*(\lambda - \lambda z)}{z^m} + \frac{h(z)z^K}{b_2^*(\lambda - \lambda z)} = 0$$

for all j for which $b^*(\lambda - \lambda z)g(1/z) = 1$.

Though the problems studied by Singh (1971) and Bagchi and Templeton (1973b) are similar, the results are not directly comparable, as has been noted in the latter paper. Dick (1970) considers an imbedded Markov chain approach to a more general system, wherein he assumes that the service does not start until there are at least a in the queue and that when the waiting room has M or more customers, the arrival rate changes from λ to λp, $0 \leqslant p \leqslant 1$.

14.4 MODELS WITH EXPONENTIAL SERVICE TIME

Of the systems with general independent input and exponential service time, two need special mention, the systems $GI/M^Y/1$ and $GI/M(a, b)/1$. Bhat (1967, 1968) and Teghem *et al.* (1969) have dealt exhaustively with $GI/M^Y/1$; the other system $GI/M(a, b)/1$ has received attention only recently. Madill (1985) has examined the system $GI/M(a, b)/1$ through the imbedded Markov chain approach. The limiting state probabilities at prearrival epochs are given in terms of the inter-arrival zero service probability, k_0 and the real root w of the characteristic equation

$$w = a^*(\mu(1 - w^b))$$

where $a^*(\cdot)$ is the L.T. of the inter-arrival time distribution. Using renewal theoretic arguments the random epoch probabilities are derived therefrom, and then the post-departure probabilities are obtained. Madill also gives expression for various measures of efficiency at each of the three epochs.

14.5 ALGORITHMIC APPROACH

Neuts (1979) introduces an algorithmic approach which avoids application of Rouché's theorem. The analytic procedure uses a matrix-geometric approach as an alternative to the classical methods, and the computational procedure is iterative. The method and its extensions have been extensively treated in his book (1981), and in an invited review paper [Neuts (1984)]. The method has been employed by several other researchers. Baba (1983) has used the matrix-geometric method for the study of bulk queues.

14.6 NUMERICAL APPROACH

It has been said that queueing theory wears a Laplacian curtain. To raise this curtain, some authors have recently attempted to either invert the transforms or produce alternative methods. Neuts, for example, (see section 5) has proposed a matrix-geometric approach. Other authors such as Powell (1981), Chaudhry *et al.* (1984), Easton and Chaudhry (1982) have been concentrating to invert the transforms by finding roots. This method seems to have worked well in a great many queueing problems. For details of the application of this method to bulk-service queueing problems, see Powell (1981), Chaudhry *et al.* (1984), Easton and Chaudhry (1982) (see also Kambo and Chaudhry (1984)), Holman *et al.* (1981), Cromie and Chaudhry (1976), Kambo and Chaudhry (1982, 1983, 1984), Madill and Chaudhry (1985), and Chaudhry, Madill and Briére (1987).

14.7 CONTROL PROBLEMS

Several authors [Kosten (1967), Deb and Serfozo (1973), Weiss (1979, 1981a, 1981b), Weiss and Pliska (1982), Medhi (1984b)] have considered control problems associated with single server queues with general bulk service rule. Such bulk service queues could be used as models of a shuttle transportation system, in which vehicles ply between two terminals say, $T1$ and $T2$. The objective function is the long run average wait or cost function where the costs involved are a round trip fixed cost K and a waiting cost per unit $h(w) = h \cdot w$ (h is a positive constant and w is the waiting time) per customer. The vehicle will wait until there are Q customers available, the capacity of the vehicle being assumed to be infinite: this general service rule (Q, ∞) will be called a Q-policy. Kosten (1967), Deb and Serfozo (1973), and Weiss (1979) consider the $M/G(Q, \infty)/1$ model. While Kosten did interesting graphical analysis, Deb and Serfozo (1973) demonstrate the existence of an optimal value Q^* of Q, i.e., the existence of a control limit Q^*-policy. Ignall and Kolesar (1974) consider the model $M/D(Q, \infty)/1$ and the policy that there are Poisson arrivals at both terminals while the vehicle is controlled at $T1$ and held there until the total number of passengers at both the terminals is Q or more. They obtain an expression for the average cost function and claim that $\sqrt{(2\lambda K/h)}$ is a good approximation for optimal Q. Weiss (1979) considers the model $M/G(Q, \infty)/1$ and shows that the average cost per unit time $C(Q)$ is given by

$$C(Q) = \frac{K + (\lambda h/2)E(D^2) + (h/2\lambda) \sum_{n=0}^{Q-1} p_n(Q^2 - Q - n^2 + n)}{E(D) + \sum_{n=0}^{Q-I} p_n(Q - n)/\lambda}$$

where, D is the round trip time, p_n is the probability that exactly n arrive during D.

With $K = 0$, $h = 1$, one finds that the average waiting time $F(Q)$ is given by

$$F(Q) = \frac{\lambda^2 E(D^2)/2 + (1/2) \sum_{n=0}^{Q-1} p_n(Q^2 - Q - n^2 + n)}{\lambda E(D) + \sum_{n=0}^{Q-1} p_n(Q - n)}.$$

Weiss finds that the value Q^* of Q which minimizes the average wait $F(Q)$ is given by

$$Q^* = \min\left\{Q \in I \mid 2QE(D) + (1/\lambda) \sum_{n=0}^{Q-1} p_n[(Q-n)(Q-n+1)] \geqslant (2k/h) + \lambda E(D^2)\right\}.$$

He also obtains an expression for the d.f. of the waiting time distribution for a shuttle system with constant trip time. In another paper Weiss (1981) further investigates the optimality criterion with regard to characterization of the optimal control limit. Powell (1981) considers a number of control policies besides the one considered above. However there are several open questions relating to control problems, which perhaps deserve greater attention such as extension to cases with finite capacity, control at both terminals, comparison of various control policies (including cancellation) and so on. See also Sim and Templeton (1982), Powell (1981, 1983), and Powell and Humblet (1986).

There are certain other areas of bulk service queues which have not received enough attention, such as approximations, conservation laws, bounds and inequalities, and so on.

Acknowledgement. The second author gratefully acknowledges the research facilities and financial support provided by the University of Toronto (through Professor Templeton) and by the Royal Military College of Canada (through Professor Chaudhry).

REFERENCES

1. Baba, Y. (1983): Algorithmic methods for *PH/PH/*1 queues with batch arrival or services, *J. Opns. Res. Soc. Japan*, **26**, 12–32.

2. Bagchi, T. F. and Templeton, J. G. C. (1973a): Finite waiting space bulk queueing systems, *J. Eng. Math.*, **7**, 313–317.

3. ——— (1973b): A note on the $M^x/G^y/1$, bulk queueing system., *J. Appl. Prob.*, **10**, 901–906.

4. Bailey, N. T. J. (1954): On queueing processes with bulk service, *J. Roy. Statist. Soc.*, **B16**, 80–87.

5. Bhat U. N. (1964): Imbedded Markov chain analysis of a single server bulk queue, *J. Austral. Math. Soc.*, **4**, 244–263.

6. ——— (1967): Some explicit results for the queue *GI/M/*1 with group service, *Sankhyā*, **A29**, 199–206.

7. ——— (1968): *A Study of the Queueing Systems M/G/*1 *and GI/M/*1, Lecture Notes in O.R. and Math. Eco., Springer-Verlag, Heidelberg.

8. Bloemena, A. B. (1960): On queueing processes with a certain type of bulk service, *Bull. Inst. Intern. Statist.*, **37**, 219–226.

9. Borthakur, A. (1971): A Poisson queue with a general bulk service rule, *J. Ass. Sc. Soc.*, **XIV**, 162–167.

10. ——— (1975): On busy period of a bulk queueing system with a general rule for bulk service, *Opsearch*, **12**, 40–46.

11. ——— (1976): *A Study of Some Queueing Models.* Ph.D. thesis, Gauhati University, India.

12. ——— (1981): Joint distribution of the number of customers in the queue and the number of batches served at time *t* within a busy period, *R.A.I.R.O.*, **15**, 271–277.

13. Borthakur, A. and Medhi, J. (1974): A queueing system with arrival and service in batches of variable size, *Cah. Cen. Etud. Rech. Oper.*, **16**, 117–126.

14. Chang, W. (1970): Single server queueing processes in computing systems, *Inter. Bus. Machines Systems, J.*, **9**, 36–71.

15. Chaudhry, M. L. and Lee, A. M. (1972): Single channel constant capacity bulk service queueing process with an intermittently available server, *INFOR*, **10**, 284–291.

16. Chaudhry, M. L., Madill, B. R. and Chaudhry, Y. (1984): Steady-state probabilities of some bulk and nonbulk queues—computer results with high parameter values. Presented at the TIMS XXVI international meeting, Denmark, 1984.

17. Chaudhry, M. L., Madill, B. R. and Brière, G. (1987): Computational Analysis of steady-state probabilities of $M/G(a, b)/1$ and related non-bulk queues, *Queueing Systems*, **2**, 93.

18. Chaudhry, M. L. and Templeton, J. G. C. (1968): On the discrete-time queue length distribution in a bulk service system considering correlated arrivals. *Canadian. Oper. Res. Soc. J.*, **6**, 79–88.

19. ——— (1981): The queueing system $M/GB/1$ and its ramifications, *Euro. J. Opns. Res.*, **6**, 56–60.

20. ——— (1983): *A First Course in Bulk Queues*, Wiley, New York.

21. Cohen, J. W. (1969): *The Single Server Queue*, North-Holland, Amsterdam (Revised Ed. 1982).

22. Cosmetatos, G. P. (1983): Increasing productivity in exponential queues by server-sharing, *Omega*, **11**, 287–193.

23. Crane, M. A. (1974): Queues in transportation II—an independently despatched system, *J. Appl. Prob.*, **11**, 145–158.

24. Craven, B. D. (1963a): Some results for the bulk service queue, *Aust. J. Statist.*, **5**, 49–56.

25. ——— (1963b): Asymptotic transient behaviour of the bulk service queue, *J. Austral. Math. Soc.*, **3**, 503–512.

26. Cromie, M. V. and Chaudhry, M. L. (1976): Analytically explicit results for the queueing system $M/M^x/C$ with charts and tables for certain measures of efficiency, *Opl. Res. Q.*, **27**, 733–745.

27. Deb, R. K. and Serfozo, R. F. (1973): Optimal control of batch service queues, *Adv. Appl. Prob.*, **5**, 340–361.

28. Dick, R. S. (1970): Some theorems on a single server queue with balking, *Operations Res.*, **18**, 1193–1205.

29. Downton, F. (1955): Waiting time in bulk service queues, *J. Roy. Statist. Soc.*, **B17**, 256–261.

30. ——— (1956): On limiting distributions arising in bulk service queues, *J. Roy. Statist. Soc.*, **B18**, 265–274.

31. Easton, G. D. and Chaudhry, M. L. (1982): The queueing system $E_k/(a, b)/1$ and its numerical analysis, *Comp. & Opns. Res.*, **9**, 197–205.

32. Fabens, A. J. (1961): The solution of queueing and inventory models by semi-Markov processes, *J. Roy. Statist. Soc.*, **B23**, 113–117.

33. Fabens, A. J. and Perera, A. G. A. D. (1963): A correction to the above, *J. Roy. Statist. Soc.*, **B25**, 455–456.

34. Finch, P. D. (1958): The effect of the size of the waiting room on a simple queue, *J. Roy. Statist. Soc.*, **B20**, 182–186.

35. ——— (1962): On the transient behaviour of a queueing system with bulk service and finite capacity, *Ann. Math. Statist.*, **33**, 973–984.

36. Foster, F. G. and Nyunt, K. M. (1961): Queues with batch departures I, *Ann. Math. Stat.*, 32, 1324–1332.

37. Foster, F. G. and Perera, A. G. A. D. (1964): Queues with batch departures II, *Ann. Math. Stat.*, **35**, 1147–1156.

38. Fujisawa, T. (1962): On a waiting-line process with a particular bulk service, *Yokohama Math. J.*, **10**, 45–51.

39. Fukuta, J. (1969): Concentrated service queue with limited unit source, *J. Opns. Res. Soc., Japan*, **12**, 21–35.

40. Ghosal, A. (1970): *Some aspects of queueing and storage systems*, Springer-Verlag, New York.

41. Goyal, J. K. (1967): Queues with hyper-Poisson arrivals and bulk exponential service, *Metrika*, **13**, 155–167.

42. Giffin, W. C. (1966): A simplified model for bulk service at a series of stations, *J. Ind. Engg.*, **17**, 430–436.

43. Gross, D. and Harris, C. M. (1985): *Fundamentals of queueing theory*, 2nd ed., Wiley, New York.

44. Harris, C. M. and Markin, P. G. (1972): A note on feedback queues with bulk service, *J. Assoc. Comp. Math.*, **19**, 727–733.

45. Holman, D. F. and Chaudhry, M. L. and Ghosal, A. (1981): Some results for the general bulk service queueing system, *Bull. Austral. Math. Soc.*, **23**, 161–179.

46. Ignall, E. and Kolesar, P. (1972): Operating characteristics of a simple shuttle under local dispatching rules, *Opns. Res.*, **20**, 1077–1088.

47. ——— (1974): Optimal dispatching of an infinite capacity shuttle: Control at a single terminal, *Opns. Res.*, **22**, 1003–1024.

48. Jaiswal, N. K. (1960a): Bulk-service queueing problem, *Opns. Res.*, **8**, 139–143.

49. ——— (1960b): Time-dependent solution of the bulk-service queueing problem, *Opns. Res.*, **8**, 773–781.

50. ——— (1961): A bulk service queueing problem with variable capacity, *J. Roy. Statist. Soc.*, **B23**, 143–148.

51. Kambo, N. S. and Chaudhry, M. L. (1982): Distribution of the busy period for the bulk service queueing system $E_k/M(a, b)/1$, *Comp. & Opns Res.*, **11**, 267–274.

52. ——— (1984a): On two bulk-service queueing models, *J. Infor. and Opt. Sci.*, **5**, 280–292.

53. ——— (1984b): On distributions of waiting and idle times of two related bulk-service queues, *Sankhyā*, **46**, Ser. B, 1–9.

54. Kashyap, B. R. K. (1966): The double-ended queue with bulk service and limited waiting space, *Opns. Res.*, **14**, 822–834.

55. Kendall, D. G. (1951): Some problems in the theory of queues, *J. Roy. Statist. Soc.*, **B13**, 151–173.

56. Kosten, L. (1967): The custodian problem, *Queueing Theory, Recent Developments and Applications* (Ed. R. Cruon), English Univ. Press, 65–70.

57. Kotiah, T. C. T., Thompson, J. W. and Waugh, W. A. O'N, (1969): Use of Erlangian distributions for single server queueing systems, *J. Appl. Prob.*, **6**, 584–593.

58. Le Gall, P. (1962): *Les systémes, avec ou sans attente et les processus stochastiques*, Dunod, Paris.

59. Luchak, G. (1958): The continuous time solution of the equations of the single channel queue with general class of service time distributions by the method of generating functions, *J. Roy. Statist. Soc.*, **B20**, 176–181.

60. Lwin, T. and Ghosal, A. (1971): On the finite capacity treatment of Bailey's queueing model, *Bull. Cal. Statist. Assoc.*, **20**, 67–76.

61. Madill, B. R. and Chaudhry, M. L. (1985a): On the queueing system $GI/M(a, b)/1$ (under submission).

62. ——— (1985b): Waiting time distributions in the queueing system $GI/M(a,b)/1$ (under submission).

63. Medhi, J. (1975): Waiting time distribution in a Poisson queue with general bulk service rule, *Management Sci.*, **21**, 277–283.

64. ——— (1979): Further results in a Poisson queue under general bulk service rule, *Cah. Cent. Etud. Rech. Oper.*, **21**, 183–189.

65. ——— (1982): *Stochastic Processes*, Halsted Wiley, New York.

66. ——— (1984a): *Recent Developments in Bulk Queueing Models*, Wiley Eastern, New Delhi.

67. ——— (1984b): Bulk service queueing models and some associated control problems. Statistics: Applications and New Directions, *Golden Jubilee Intl. Conference Volume*, Indian Statistical Institute, Calcutta, 369–377.

68. Mercer, A. (1968): A queue with random arrivals and scheduled bulk departures, *J. Roy. Statist. Soc.*, **B30**, 185–189.

69. Mohan, C. and Murari, K. (1972): Time dependent solution of a correlated queueing problem with variable capacity, *Metrika*, **19**, 209–215.

70. Mohanty, S. G. (1972): On queues involving batches, *J. Appl. Prob.*, **9**, 430–435.

71. Murari, K. (1968): An additional special channel limited space queueing problem with service in batches of variable size, *Opns. Res.*, **16**, 83–90.

72. ——— (1969): A queueing problem with correlated arrivals and general service time distribution. *Z. Angew. Math. Mech.*, **49**, 151–156.

73. Nadarajan, R. and Sankaranarayanan, G. (1980): A bulk service queueing system with Erlang input, *J. Ind. Stat. Ass.*, **18**, 109–116.

74. Nadarajan, R. and Subramanian, A. (1984): A general bulk service queue with server's vacation, *Operational Research in Managerial Systems* (Ed. Aggarwal, S. P.), Academic Publi., New Delhi, 127–135.

75. Nair, S. S. and Neuts, M. F. (1972): Distribution of occupation time and virtual waiting time of a general class of bulk queues, *Sankhyā Ser. A*, **34**, 17–22.

76. Natarajan, R. (1962): Discrete-time bulk service queueing process, *Defence Sci. J.*, **12**, 318–326.

77. Neuts, M. F. (1966a): The busy period of a queue with batch service, *Opns. Res.*, **13**, 815–819.

78. ——— (1966b): Semi-Markov analysis of a bulk queue, *Bull. Soc. Math. Belg.*, **18**, 28–42.

79. ——— (1967): A general class of bulk queues with Poisson input, *Ann. Math. Statist.*, **38**, 759–770.

80. ——— (1979): Queues solvable without Rouché's theorem, *Opns. Res.*, **27**, 767–781.

81. ——— (1981): *Matrix-Geometric solutions in stochastic models—an algorithmic approach*, The Johns Hopkins Univ. Press, Baltimore.

82. ——— (1984): Matrix-analytic methods in queueing theory, *Euro. J. Opl. Res.*, **13**, 2–12.

83. Newell, G. F. (1960): Queues for a fixed-cycle traffic light, *Ann. Math. Statist.*, **21**, 589–597.

84. ——— (1965): Approximation methods for queues with application to the fixed-cycle traffic light, *SIAM Rev.*, **7**, 223–239.

85. Novaes, A. and Frankel, E. (1966): A queueing model for unitized cargo generation, *Operations Res.*, **14**, 100–132.

86. Powell, W. B. (1981): *Stochastic delays in transportation terminals: New results in the theory and application of bulk queues*, Ph.D. Thesis, Massachussetts Institute of Technology.

87. ——— (1983): Bulk service queues with deviations from departure schedules: The problem of correlated headways, *Transpor. Res.*, **17B**, 221–232.

88. Powell, W. B. and Humblet, P. (1986): The bulk service queue with a general control strategy: theoretical analysis and a new computational procedure, *Operations Rev.*, **34**, 267–275.

89. Prabhu, N. U. (1960): Some results for the queue with Poisson arrivals, *J. Roy. Statist. Soc.*, **B22**, 104-107.

90. ——— (1965): *Queues and inventories—A study of their basic stochastic processes*, Wiley, New York.

91. ——— (1980): *Stochastic Storage Processes, Queues, Insurance Risks and Dams*, Springer-Verlag, New York.

92. Sim, S. H. and Templeton, J. G. C. (1982): Queue-dependent vehicle dispatching with options and optimal fleet size, *Transpn. Res.*, **16B**, 233–243.

93. Singh, V. P. (1971): Finite waiting space bulk service system, *J. Engg. Math.*, **5**, 241–248.

94. ——— (1972): Addendum to the above paper, *J. Engg. Math.*, **6**, 85-86.

95. Smith, W. L. (1953): On the distribution of queueing times, *Proc. Cambridge Philos. Soc.*, **49**, 449–461.

96. Stidham, S. (1974): Stochastic clearing systems, *Stochastic Proc. and their Applications*, **2**, 85–113.

97. Stuart, I. M. and McMahon, G. B. (1966): A queueing system with bulk service, *Operations Res.*, **14**, 728–731.

98. Takács, L. (1965): Investigation of waiting time problems by reduction to Markov processes, *Acta Math. Acad. Sci. Hungar.*, **6**, 101–129.

15

Stochastic Models in Software Reliability Based on Non-Homogeneous Poisson Process

P. K. Kapur **Shakti Bhushan**

Department of Operational Research, University of Delhi
Delhi, India

and

M. Meenakshi Bai

Department of Mathematics & Statistics, S.V. Ag. College
A.P.A. University, Tirupati, India

15.1 INTRODUCTION

As softwares form an important part of many critical missions such as space shuttles, and important systems such as nuclear reactors and heart monitors, the reliable operation of these projects/systems depends critically on the reliable operation of their software components. Because of total dependence of these systems on the underlying software, the concept of Software Reliability has gained considerable importance.

The life cycle (LC) of software involves a series of production activities and can generally be divided into four phases: design, coding, testing and operation/maintenance[4]. In spite of great advancement of the programming technology, the chances of error/fault occurrence due to human imperfection at every step are many. In other words, software can never be made error/fault/bug free. A fault/error is the defect in the software that, when executed under particular conditions, causes a failure, where failure means that the program in its functioning has not met user's requirements in some way. To remove these faults/errors the software is tested under a large number of representative test cases with the intention of exposing faults/errors possibly contained in the program. When a failure is observed, the code is inspected to find the fault/error which caused the failure and an attempt is made to fix it. As a result reliability of the software is expected to increase. Thus it is very important to know the number of errors/faults remaining in the software or time interval between software failures. One of the approaches to software reliability is to describe a software error detection process which represents the behaviour of errors/faults during testing phase of the software development. Many models have been developed which attempt to estimate the error content of a software, and to predict the software reliability. These models are called 'Software Reliability Growth Models' (SRGM) since due to testing of a software (and hence removing faults/errors) its reliability is expected to grow.

In this paper we summarise and review models based on Non-Homogeneous Poisson Process (NHPP). Models discussed are either based on failure intensity function or the mean value function. The two are not necessarily interchangable. Besides, it is also of utmost importance to decide the release time of the software. Several methods have been suggested in the literature. In this paper we also give a brief overview of the release policies.

15.2 ASSUMPTIONS

Some of the general assumptions (apart from some special ones for specific models discussed) assumed in every model are as follows:

1. Software system is subject to failure during execution caused by errors/faults remaining in the system.
2. Failure rate of the software is equally affected by errors/faults remaining in the software.
3. The number of failures detected at any time is proportional to the remaining number of errors/faults in the software.
4. On a failure repair effort starts and fault/error causing failure is removed with certainty.
5. All faults/errors are mutually independent from failure detection point of view.
6. The proportionality of failure detection/fault isolation/fault removal is constant.
7. Corresponding to the error detection/removal phenomenon at the manufacturer/user end, there exists an equivalent error detection/removal phenomenon at the user/manufacturer end.
8. Software life cycle is more than the optimal release time.
9. The error detection/removal phenomenon is modelled by NHPP.

15.3 NOTATIONS

$[N(t); t > 0]$	Counting process representing the cumulative number of failures/isolations/removals in (o, t).
a	*S*-expected initial error/fault content.
b	Error detection/isolation/removal rate per error.
b_i/b_f	Initial/Final value of b.

$m_f(t)/m_i(t)/m_r(t)$	*S*-expected number of failures/isolations/removals in (o, t).
$b(m_f)/b(t)$	Failure/error-detection rate per error as a function of $m_f(t)/t$.
$\lambda(t)$	Failure intensity at t.
$w(t)$	Current testing effort expenditure at time t.
$W(t)$	Cumulative testing effort by time $t = \int_0^t w(x)\, dx$.
α, β	Parameters of testing effort function.
$R(x \mid t)$	Reliability of a software in $(t, t + x)$.
λ_0	Initial failure intensity.
T_{1c}	Life Cycle of the Software.
C_1/C_2	Cost of removing a fault/error before/(after) releasing.
C_3	Testing Cost Per unit time.
$C(T)$	Total Cost function.
λ_d/R_d	Desired level of failure intensity (reliability) to be achieved.

15.4 SOME DEFINITIONS

Non-Homogeneous Poisson Process (7, 28)

Let $(N(t); t \geqslant 0)$ be a counting process representing the cumulative number of failures (or faults isolated or removed as the case may be) by time t. $N(t)$ is a random variable and $N(t)$ process is NHPP if

(a) $N(0) = 0$

(b) $(N(t); t \geqslant 0)$ has independent increments.

(c) Pr(two or more events in $(t, t + h)) = o(h)$

(d) Pr[exactly one event in $(t, t + h)] = \lambda(t)\cdot h + o(h)$

where $\lambda(t)$ is intensity function of $N(t)$.

And if we let

$$m(t) = \int_0^t \lambda(x)\, dx$$

represent $m_f(t)$ or $m_i(t)$ or $m_r(t)$ depending upon the model, then it can be shown that

$$\Pr[N(t) = n] = \frac{[m(t)]^n e^{-m(t)}}{n!}$$

$$n = 0, 1, 2 \ldots$$

i.e. $N(t)$ has Poisson distribution with expected value $E[N(t)] = m(t)$ for $t > 0$ and $m(t)$ is called the mean value function of the NHPP.

****Software Reliability Growth Model (29)**

Software reliability growth model is defined as a mathematical relation between the time span of the testing (or using) the software and the cumulative number of errors/faults detected.

Software Reliability (9)

Software reliability is defined as the probability that a software failure does not occur in $(t, t+x]$ given that the most recent failure occurred at time $t > 0, x > 0$. It can be shown that for a model based on NHPP.

$$R(x/t) = \exp(-[m_f(t + x) - m_f(t)]).$$

Failure Intensity (24)

The failure intensity function is the state of change of the mean value function representing the average cumulative number of failures associated with the given time point, i.e.

$$\lambda(t) = \frac{d}{dt}[m_f(t)].$$

It may also be defined as the number of failures per unit time.

A. MODELS BASED ON FAILURE INTENSITY

Starting with the failure intensity function, Musa proposed two models (Second with Okumoto) through which it is attempted to predict future failure behaviour and reliability of a software.

A.1 Basic Execution Time Model [Musa (25)]

It has a failure intensity function which decays exponentially with execution time t, i.e.

$$\lambda(t) = \lambda_0 e^{-bt} \qquad \text{(A.1.1)}$$

where b is the rate of decrease per unit time (which is same as error detection rate per error).

The initial failure intensity λ_0 is given as

$$\lambda_0 = ab. \qquad \text{(A.1.2)}$$

The expression for the expected number of failures by time t can be obtained from (A.1.1) as

$$m_f(t) = a(1 - e^{-bt}) \qquad \text{(A.1.3)}$$

And the failure intensity can be expressed as a function of $m_f(t)$ as

$$\lambda(m_f) = b[a - m_f(t)] \qquad \text{(A.1.4)}$$

(A.1.4) shows that the rate of change of failure intensity with respect to failures experienced is constant whether it is the first failure or the last that is being fixed.

A.2 Logarithmic Poisson Model [Musa et al. (26)]

This model has a failure intensity function which decays exponentially with respect to the mean value function $m_f(t)$, i.e.

$$\lambda(t) = \lambda_0 e^{-bm_f(t)} \tag{A.2.1}$$

where b, in this case, is rate of decrease per failure.

From (A.2.1), we get

$$m_f(t) = \ln\,[1 + \lambda_0 bt]/b \tag{A.2.2}$$

The name "Logarithmic Poisson Model" is derived from the form of (A.2.2).

The failure intensity as a function of $m_f(t)$ is given as

$$\lambda(m_f) = \lambda_0 [e^{-b}] m_f(t) \tag{A.2.3}$$

This shows that the rate of change of intensity with respect to failures experienced, decreases exponentially with failures experienced and is not constant as in the case of basic execution time model. It means that the first failure initiates a repair process that yields a substantial decrease in failure intensity, while later failures result in much smaller decrements. This is because during testing, the software is first tested on frequently used inputs.

Moreover, it may be noted that by time infinity, the failure intensity reduces to zero and the number of failures experienced is infinity. This is possible when, either at the time of debugging faults are being introduced, or the debugging process is imperfect (which is discussed in the next section) or each fault generates more than one failure (which is also discussed later in the paper) or, when combinations of these are possible. It is obvious in this case that the parameters of interest are not the number of errors (which can be infinite) in a software, but the failure intensity and the rate at which failures are occurring. As has been pointed out by Musa this model is applicable under highly non-uniform operational-profile.

B. EXPONENTIAL MODELS BASED ON MEAN VALUE FUNCTION

This category of SRGMs is based on slightly different assumption. Models are formulated based on an expression for the mean value function of the Poisson process rather than the failure intensity function and are suitable for finite error content.

B.1. Goel-Okumuto Model [Goel et al. (9)]

Assuming that the S-expected number of failures in $(t, t + \Delta t)$ is essentially proportional to the S-expected number of undetected errors at time t (which is the underlying assumption in all these models), following equation can be easily written.

$$m_f(t + \Delta t) - m_f(t) = b[a - m_f(t)]\Delta t + o(\Delta t) \tag{B.1.1}$$

where $o(\Delta t)/\Delta t \to 0$ as $\Delta t \to 0$. By letting $\Delta t \to 0$, (B.1.1) gives

$$m_f'(t) = b[a - m_f(t)] \qquad \text{(B.1.2)}$$

(B.1.2), together with $m_f(o) = 0$, gives

$$m_f(t) = a(1 - e^{-bt}) \qquad \text{(B.1.3)}$$

which is mathematically isomorphic to the mean value function of Basic execution time model. Here it is assumed that the fault causing failure is immediately removed and hence fault detection and failure occurrence are assumed to be synonymous.

B.2. Modified Exponential Model [Yamada et al. (38)]

In general the chance of detecting an error on a given test run is not constant. There are some failures which are easy to detect whereas there are some which are difficult to detect (38). Similarly, errors can be classified in terms of their effect on the output as critical, major and minor (23).

Here we describe a SRGM considering two types of errors, type 1 and type 2.

Suppose that p_i $(i = 1, 2)$ is the error content proposition of type i $(i = 1.2)$ error, where $p_1 + p_2 = 1$. Then, the initial error content of type i $(i = 1.2)$ error is $p_i a$.

Thus, we have

$$m_{if}(t) = ap_i[1 - e^{-b_i t}]\ i = 1, 2 \qquad \text{(B.2.1)}$$

where $m_{if}(t)$ is the expected number of failures due to type i $(i = 1, 2)$ error, b_i is error detection rate per type i $(i = 1, 2)$ errors.

So

$$m_f(t) = \sum_i ap_i[1 - e^{-b_i t}] \qquad \text{(B.2.2)}$$

B.3. Exponential Model with Imperfect Debugging [Kapur et al. (14)]

During testing (or operational phase till the software is supported) phase of a software, on a failure an attempt is made to correct the cause of the failure. However, it is not always possible to find the cause of the failure and remove it. This may be attributed to lack of sufficient knowledge about the software, poor documentation of the software and so on.

In this model it is assumed that on a failure instantaneous repair effort starts and the following may occur:

(a) fault content is reduced by one with probability p_0,
(b) fault content is unchanged, with probability $1 - p_0$.

Based on these assumptions, following differential equation may easily be written

$$m_r'(t) = bp_0[1 - m_r(t)] \qquad \text{(B.3.1)}$$

This gives

$$m_r(t) = a[1 - e^{-bp_0t}] \tag{B.3.2}$$

and

$$m_f(t) = \frac{a}{p_0}[1 - e^{-bp_0t}] \tag{B.3.3}$$

(B.3.3) implies that the expected number of failures by infinite time will be greater than 'a' which is because that some faults may not be removed even though an attempt was made to do it.

B.4. Exponential Model with Testing Effort [Yamada et al. (37)]

In general some resources like, manpower C.P.U. time etc. are spent during testing phase of the software development. The consumption curve of testing resources over the testing period can be thought of as a testing-effort curve.

In this model it is assumed that the *S*-expected number of errors detected in the time interval $(t, t + \Delta t)$ to the current testing effort expenditure is proportional to the *S*-expected number of remaining errors. So

$$\frac{m_f'(t)}{w(t)} = b[a - m_f(t)] \tag{B.4.1}$$

This gives

$$m_f(t) = a[1 - e^{-bw(t)}] \tag{B.4.2}$$

It is generally assumed that testing effort in the software development process follows exponential or Rayleigh curve, i.e.

$$W(t) = \alpha[1 - e^{-\beta t}] \tag{B.4.3}$$

or

$$W(t) = \alpha[1 - e^{-\beta t^2/2}] \tag{B.4.4}$$

Values of α and β are estimated respectively from (B.4.3) (or B.4.4). These estimated values of α and β are then put into (B.4.2) to get the estimates of a and b.

Since resources are finite, $ae^{-\alpha b}$ number of faults will remain undetected by time infinity.

B.5. Flexible Model [Bittanti et al. (2)]

Instead of having a constant error detection rate per error throughout the testing phase, here it is assumed that it is a function of number of errors detected at that time, i.e.,

$$b(m_f) = b_i + \frac{b_f - b_i}{a} m_f(t) \tag{B.5.1}$$

And so, we have

$$m_f'(t) = b(m_f)[a - m_f(t)] \tag{B.5.2}$$

From (B.5.2), we get

$$m_f(t) = \frac{ab_i[e^{b_f t} - 1]}{b_f + b_i[e^{b_f t} - 1]} \tag{B.5.3}$$

Depending upon b_i and b_f this model depicts different reliability growth phenomenon. It is in this sense the model is flexible. Different cases of the model are discussed below:

(i) Constant error detection rate per error, i.e. $b_i = b_f = b$. Then (B.5.3) reduces to Goel-Okumoto model (9).

(ii) Decreasing error detection rate per error. In this case the average failure growth curve is no more exponential and lies in the region delimited by the exponential curves with $b = b_f$ and $b = b_i$ respectively.

(iii) Vanishing error detection rate per error, i.e., $b_i > 0$, $b_f = 0$. Then (B.5.3) reduces to

$$m_f'(t) = \frac{b_i[a - m_f(t)]^2}{a} \tag{B.5.4}$$

This gives

$$m_f(t) = \frac{ab_i t}{1 + b_i t} \tag{B.5.5}$$

Expression similar to (B.5.4) is also suggested by Min Xie (31) in a different context.

(iv) Increasing error detection rate per error, i.e., $b_f > b_i$. It can be seen that when $b_f > 2b_i$, then at

$$t = \frac{1}{b_f} \ln\left[\frac{b_f - b_i}{b_i}\right] \tag{B.5.6}$$

The curve for the average number of failures has a point of inflexion and the reliability growth phenomenon is called *S*-shaped which is discussed in greater detail in the next section.

B.6. Model for an Error Removal Phenomenon [Kapur et al. (12)]

It is a model with the assumption that during the detection/removal of a fault some proportion of the remaining faults may also be detected/removed without these faults causing any failure.

It is assumed, for this model, that c is the error detection/removal rate of additional errors detected/removed, then we get the following differential equation:

$$m_r'(t) = b[a - m_r(t)] + \frac{a - m_r(t)}{a} cm_r(t) \tag{B.6.1}$$

Solving it, we get

$$m_r(t) = ab\,\frac{e^{(c+b)t} - 1}{c + be^{(c+b)t}} \tag{B.6.2}$$

and

$$m_f(t) = \frac{ab}{c} \ln \left[\frac{b + c}{b + ce^{-(b+c)t}} \right] \tag{B.6.3}$$

In this model number of failures by infinity will be less than 'a' since we will remove these many faults/errors much before 'a' number of failures.

B.7. A General SRGM [Kapur et al (16)]

Assuming a time dependent failure rate per error, a general form of the mean value function is given by

$$m_f(t) = a[1 - \exp\left\{- \int_0^t b(x)dx\right\} \tag{B.7.1}$$

and the failure intensity is given by

$$\lambda(t) = ab(t) \exp\left\{- \int_0^t b(x)dx\right\} \tag{B.7.2}$$

Depending on the nature of $b(t)$, several SRGMS can be obtained.

C. *S*-SHAPED SOFTWARE RELIABILITY GROWTH PHENOMENON

It is generally accepted that there exists *S*-shaped software reliability growth phenomenon that is observed in the testing phase of the development of the softwares. It has been interpreted in different ways. Ohba (27) interprets it as initial delay of fault isolation after the initial failure detection. Yamada (39) interprets it as unskilledness of the testing team. Ohba (27) developed a SRGM for a situation where a fault in a software is dependent on the previous fault detected and the cumulative number of failures/faults detected curve of this model also shows *S*-shaped phenomenon. This same *S*-shaped phenomenon is again observed when the testing effort curve of Yamada *et. al.* (37) follow Rayleigh type of distribution. Again in flexible model of Bittanti (2) this phenomenon is observed when the error detection rate per error is increasing.

C.1. Delayed *S*-Shaped SRGM [Ohba (27)]

Fault removal, in this model, is assumed to be a two phase process consisting of failure detection and its eventual removal by isolation. It takes care of the time taken to isolate and remove a fault and so it is important that the data to be used here should be that of fault isolation.

It is further assumed that the number of faults isolated at any time is proportional to the current number of faults not isolated. Failure rate and isolation rate per error are assumed to be same and equal to b.

Thus

$$m_f'(t) = b[a - m_f(t)] \tag{C.1.1}$$

$$m_r'(t) = b[m_f(t) - m_r(t)] \tag{C.1.2}$$

Solving these, we get the mean value function as

$$m_r(t) = a\{1 - (1 + bt)e^{-bt}\} \tag{C.1.3}$$

This model is called S-shaped because the graph of cumulative number of faults removed with respect to time has a lump at the initial portion of the graph.

This model can further be extended depending upon the severity of the error in the software.

C.2 Inflection S-Shaped SRGM [Obha (27)]

The mean value function of this model is given by

$$m_f(t) = \frac{1 - e^{-bt}}{1 + \phi e^{-bt}} \tag{C.2.1}$$

where ϕ is called inflexion parameter and is equal to $\frac{1-r}{r}$

where r is the ratio of the number of detectable faults to the total number of faults.

This model takes care of the situation where errors/faults are mutually dependent and is based on the assumption that the error detection rate per error increases throughout the test period.

C.3. Flexible SGRM [Bittanti et al. (2)]

As discussed in B.5 of the last section, mean value function of this model is given by

$$m_f(t) = \frac{ab_i[e^{(b_f t)} - 1]}{b_f + b_i[e^{(b_f t)} - 1]} \tag{C.3.1}$$

When $b_f > 2b_i$ then the reliability growth phenomenon is S-shaped.

C.4. Exponential with Rayleigh Type Testing Effort [Yamada et al. (37)]

As discussed in B.4, its mean value function is given by

$$m_f(t) = a[1 - e^{-bw(t)}] \tag{C.4.1}$$

And when the testing effort curve follows a Rayleigh Curve given by

$$W(t) = \alpha[1 - e - \beta^{t^2/2}] \tag{C.4.2}$$

Then the reliability growth phenomenon is S-shaped.

C.5. General S-Shaped SRGM [Kapur et al. (17)]

Error removal in this case is a three phase phenomenon consisting of fault detection, fault isolation and then its eventual removal. Rate at each phase is assumed to be equal to b.

The mean value function of this model is given by

$$m_r(t) = a\left[1 - \left(1 + bt + \frac{b^2t^2}{2}\right)e^{-bt}\right] \tag{C.5.1}$$

This model is suitable for the fault removal data.

C.6. Rayleigh Type SRGM [Kapur et al. (11)]

There is yet another software reliability growth model, which describes a S-shaped growth phenomenon. Its mean value function is given by

$$m_f(t) = a[1 - e^{-bt^2/2}] \tag{C.6.1}$$

Comparison of this model in terms of their predictive validity is given in (11).2.

D. DISCRETE SOFTWARE RELIABILITY GROWTH MODELS

In this class of SRGMs the number of test runs or the number of executed test cases is taken as the unit of error detection period. Random variable of the NHPP is defined as the number of errors detected (removed) by n tests runs.

For this class of SRGMS, assumptions are modified in the discrete sense. Little work has been done in this class of SRGMs.

D.1. Error-Content Proportional Detection Rate Model [Yamada et al. (35)]

It is assumed that the expected number of failures between nth and $(n+1)$th executed test cases is proportional to the expected number of faults remaining in it, i.e.

$$m_f(n+1) - m_f(n) = b[a - m_f(n)] \tag{D.1.1}$$

Solving this difference equation, we get

$$m_f(n) = a[1 - (1 - b)^n] \tag{D.1.2}$$

and

$$\lambda(n) = ab(1 - b)^n \tag{D.1.3}$$

D.2. Geometric Error Detection Rate SRGM (Yamada et al. (35)]

This SRGM is developed under the assumption that the ratio of the error detection rate for any test run and the rate for its predecessor is constant less than unity, and the expected number of failures per test run is geometrically decreasing. So

$$m_f(n+1) - m_f(n) = Dr^n \tag{D.2.1}$$

where r is the decreasing ratio for the expected number of error detected per test run and D is the initial expected number of errors detected by the first test run.

Thus

$$m_f(n) = D\frac{1 - r^n}{1 - r} \tag{D.2.2}$$

And the expected number of errors to be eventually detected i.e. the expected initial error content is given by

$$m_f(\propto) = D/(1 - r) \tag{D.2.3}$$

D.3. Discrete *S*-Shaped SRGM [Kapur et al. (13)]

This model takes care of the delay between the failure caused by a fault and its subsequent detection. It is a two phase phenomenon and so the mean value function of the model is given by

$$m_r(n) = a[1 - (1 + bn)(1 - b)^n] \tag{D.3.1}$$

E. ESTIMATION OF PARAMETERS

Our main aim in software reliability modelling is to estimate different model parameters so that the future behaviour of the software can be predicted. Suppose that the data on *n* failure times (or isolation or removal times) $S(s_1, s_2, \ldots, s_n)$ where $(s_1 < s_2 < \ldots s_n)$ are observed during testing. Then the likelihood function for unknown parameters of the model given *S* is given by (8):

$$L = e^{-m(S_n)} \underset{i}{\pi} \epsilon(S_i) \tag{E.1}$$

where

$$\epsilon(S_i) = \frac{d}{dt} m(t) \text{ at } t = S_i \tag{E.2}$$

and $m(t)$ is the mean value function of the underlying NHDP.

From (E.1), maximum likelihood estimates of parameters can easily be obtained.

Alternatively, suppose that the data on cumulative number of failures (or isolations or removals) $Y_k(0 < y_1 < y_2 \ldots < y_n)$ in a given time interval $(0, t_k]$ are observed. Then the likelihood function of the unknown parameters is given by (37):

$$L = \underset{k}{\pi} \frac{[m(t_k) - m(t_{k-1})]^{(Y_k - Y_k - 1)}}{[Y_k - Y_{k-1}]} \exp[-(m(t_k) - m(t_{k-1})] \tag{E.3}$$

From (E.3), estimates of unknown parameters can easily be obtained.

F. PREDICTIVE VALIDITY OF MODELS [Musa et al. (24)]

It is the ability of the model to determine future failure/removal behaviour during either the test or the operational phase from present and past failure/removal behaviour in the respective phase. It is a very effective tool to compare the applicability of models on a particular data.

Here we attempt to predict the number of failures/removals that will be experienced by the end of the period of testing over which the data has been collected and compared this with actual values. Assume that n failures/removals has been observed by the end of the time S_n, where S_i is the time to ith failure/removal. Clearly $0 < S_1 < S_2 \ldots < S_n$. The failure/removal data upto time S_t $(< S_n)$ is used to estimate the parameters of the meanvalue function $m_f(t)$ or $m_r(t)$. Then the number of failures/removals by time S_n can be predicted by substituting the estimates of parameters in $m_f(t)$ or $m_r(t)$ which is compared with the actually observed number n. This is repeated for various values of S_t.

The predictive validity can be checked visually by plotting normalised relative number of errors or removals $[m_f(S_n) - n]/n$ or $[m_r(S_n) - n]/n$ against the normalised time S_t/S_n. The error/removal will approach to zero as S_t approaches S_n. If the points are positive (negative) the model tends to overestimate (underestimate). Number closer to zero implies more accurate prediction and hence the better model.

G. OPTIMAL RELEASE POLICY

It is of utmost importance to find the appropriate release time of the software. If the release of the software is unduly delayed, manufacturer may suffer in terms of penalties and revenue loss, while a premature release may cost heavily in terms of the fixes to be done after release and may even harm manufacturer's reputation. Therefore, manufacturer must have some idea about the possible attributes of the software like its initial error content, failure rate, reliability at time t and its potential release time. Thus it is of interest to know when to stop testing and release the software. Software release time problem has been classified in different ways. One is, when to release a software so that the cost incurred during the life cycle of the software, i.e. during development and operational phase is minimized or the reliability reaches a desired level. Stopping rules in this regard have been studied by many authors[6,28,31]. Lately, a new criterion for software release has been suggested by Yamada *et al.*[33] when both cost and reliability/intensity are of interest. It discusses the release policies which minimize the cost during the life cycle of the software subject to reliability/intensity reaching a desired level. Release policies for several extended models can be found in[14,16,17,18,19,20,21].

Software cost includes the cost of removing errors both before and after release (till it is supported in the field) of the software and testing cost during testing phase. Assuming that the software is released at time T, cost function in general can be written as

$$C(T) = C_1 m_r(T) + C_2[m_r(T_{1c}) - m_r(T)] + C_3 T \quad \text{(G.1)}$$

Thus the problem of finding optimal release time reduces to

$$\min C(T)$$

Subject to

$$R(x/T) > R_d \quad \text{or} \quad \lambda(T) < \lambda_d \quad \text{(G.2)}$$

$$T > 0$$

For the discrete models, T is replaced by n, the number of executed test cases.

Alternatively this problem can be redefined in terms of maximizing gain, which is defined as the difference in cost incurred when all the errors are removed during operational phase as against the cost, when some errors are removed during the testing phase and the others are removed during the operational phase. Thus denoting it by $G(T)$, we have

$$G(T) = (C_2 - C_1)m_r(T) - C_3 T \tag{G.3}$$

So, we have

$$\max G(T)$$

Subject to

$$R(x/T) > R_d \quad \text{or} \quad \lambda(T) < \lambda_d \tag{G.4}$$

$$T > 0$$

It can be seen that maximizing gain is same as minimizing cost[14,18].

We have so far assumed that the software life cycle is constant. However, in real situation it might be more appropriate to assume it to be random variable, because software system may be abandoned because new versions are available. Recently Yun & Bai[40] discussed a software release policy maximizing profit alone, when the software life cycle is random. Profit is determined based upon the price of the software and the expenditure incurred on testing, correcting an error during testing and operation. One may also incorporate the idea of penalty cost which is incurred by the manufacturer by not delivering the software by scheduled delivering time[5].

H. FURTHER RESULTS

H.1.

Musa's Log-Poisson model assumes that finite failure intensity may result in infinite number of failures. One reason being a fault may give rise to more than one failure, thus leading to infinite failures. However, one may well assume that there is an upper bound on the number of failures experienced. Incorporating this simple idea, a new mean value function can be developed which is given by

$$m_f(t) = M[1 - e^{-bat/M}] \tag{H.1}$$

where a is the initial error content resulting in M number of failures $(M > a)$, $a = M$ will result in Goel–Okumoto model.

This model has been tried by Kapur *et al.* [11*a*] on the data due to (8) and the applicability of the model has been shown.

H.2.

One of the most difficult problems in analysing a software failure/detection data is the lack of simple model validation and parameter estimation tools. Usually a complicated parameter estimation is carried out before a data is fitted using the estimated parameters. The judgement of the fitness of a model is not obvious. *Min Xie et al.*[31] recently have developed some simple models on the lines of *Duane*[5] which provide a simple graphical technique for practical applications. Mean value function of these models by simple transformation can be transformed to a linear relation as in the case for the Duane model. The plot of the cumulative number of failures on running time on a suitable scale can be fitted by a straight line. The model parameters can thus be estimated using the slope of the fitted line and its vertical intercept. As claimed by *Min Xie et al.*[31] most of the existing models based on NHPP, although have other interpretations, do not have this simple property. There is a lack of visual interpretation. Therefore, the statistical tools and numerical algorithms for parameter estimation and model validation have to be used.

In fact their procedure is based on First-Model-Validation-Then-Parameter-Estimation approach. To illustrate this procedure, we give following simple model called 'Logarithmic Law Model' which has some similarity with Log-Poisson model of Musa[26]. The mean value function of this model is given by

$$m(t) = \ln [1 + at^b] \tag{H.2}$$

For this model, when we plot the cumulative number of failures $m(t)$ on the scale $\ln [e^{m(t)} - 1]$ against $\ln (t)$, we should get a straight line fit, if the model is valid for this particular data or else we reject the model.

15.5 CONCLUSION

We have, in this paper, discussed several SRGMs based either on the failure intensity function or the mean value function of the Poisson process. The two widely used models namely Goel-Okumoto model and Musa's basic execution time model are seen to be isomorphic if the failure correction is instantaneous. Most of the models assume that the finite error content results in those many failures. In other words, there is one-to-one correspondence between the error and the failure resulting therein. Besides, the underlying assumption in these models is that the number of failure/error detected is proportional to the error remaining in the software. If one wishes to deviate from this, it is interesting to see that one must resort to modelling based on failure intensity as in the case of Logarithmic Poisson model of Musa. Interestingly in this model a finite failure intensity results in eventual infinite failures. Hence the parameter of importance is not initial error content but initial failure intensity. Most of the models that have been developed are based on the mean value function rather than the failure intensity function. Recently *Min Xie, et al.*[31] have developed simple models which have graphical representation. These are all infinite failure models. Parameters of the models are not estimated through the usual estimation techniques but rather from the slope of the graphs and their vertical intercepts.

Accuracy of the prediction of the parameters remains a problem. Besides, there is a dearth of suitable failure data-bases. Most of the models have been developed using the data available in[8,23,24]. There is a need to strengthen this aspect.

How to choose a model from among the large number of models that are available, is also of paramount importance. Some guidelines in this regard are provided in[1,3,8,10].

In conclusion, we may mention that the models described in this paper may not be all that comprehensive in the area but hopefully these are expected to provide certain directions as regards the State of the Art.

ACKNOWLEDGEMENTS

The authors are thankful to the referee for his detailed comments and suggestions. Grateful thanks are also due to Dr. R. B. Garg and Mr. Said for their helpful suggestions in the preparation of this paper.

REFERENCES

1. Bendell, A. and Nellor, P. (1986): *Software Reliability: State of Art Report*, Infotech Limited.
2. Bittanti, S., Bolzern, P., Pedrotti, E., Pozzi, M. and Scattolini, R. (1988): A flexible modelling approach for software reliability growth, in *Software Reliability Modelling and Identification*, (Ed.), Goos, G. and Hartmanis, J., Springer-Verlag, Berlin, pp. 101–139.
3. Brown, D. E. (1987): A method of obtaining software reliability measurement during development, *IEEE Trans. Reliab.*, **36**, 573–580.
4. Conte, Dunsmore and Shen (1986): *Software Engineering: Metrics and Models*, Benjamin-Cummings, Menlo-Park, C. A.
5. Duane, J. T. (1964): Learning curve approach to reliability monitoring, *IEEE Trans. Aerospace*, **AS-2**, 563–566.
6. Forman, E. H. and Singpurwalla N. D. (Dec. 1977): An empirical stopping rule for debugging and testing computer software, *Journ. of the American Stat. Assoc.*, **72**, 750–757.
7. Goel, A. L. (1980): Software error detection model with applications, *Journ. Sys. Software*, **1**, 243–249.
8. Goel, A. L. (1985): Software reliability models: assumptions, limitations and applicability, *IEEE Trans. Software Eng.*, **SE-11**, 1411–1423.
9. Goel, A. L. and Okumoto, K. (1979): Time dependent error detection rate model for software reliability and other performance measures, *IEEE Trans. Reliab.*, **28**(3), 206–211.
10. Kamuro, Y. (1987): Evaluation of software products testing phase application to software reliability growth models, in *Reliability Theory and Application*, (Ed) Osaki, S. and Cao, J., World Scientific, pp. 188–194.
11. Kapur, P. K., Bhushan, S. and Agarwala, S. (1991): On *S*-shaped reliability growth phenomenon, *Journ. Sys. Science* (submitted).

11a. Kapur, P. K., Bhushan, S. and Younes, S.: An exponential SRGM with an upper bound on the number of failures (communicated).

12. Kapur, P. K. and Garg, R. B. (1991): A software reliability growth model for an error removal phenomenon (To be published in *Software Engg. Journ.*)

13. ——— (1990): A discrete *S*-shaped software reliability growth model, *Journ. Sys. Software* (submitted).

14. ——— (1990): Optimal software release policies for software reliability growth model under imperfect debugging *R.A.I.R.O.*

15. Kapur, P. K., Garg, R. B. and Bhalla, V. K.: Release policies with random software life time and penalty cost (communicated).

16. Kapur, P. K. and Garg, R. B. (1989): A general theoretical framework for software release policies, *IEEE Trans. Reliab.* (submitted).

17. ———: A generalised *S*-shaped software reliability growth model (communicated).

18. ——— (1990): Cost reliability optimum release policies for a software system with testing effort, *OPSEARCH*, India, **27**(2).

19. ——— (1990): Optimal release policies for inflexion *S*-shaped SRGM, *Microelectron Reliability.*

20. Kapur, P. K. and Bhalla, V. K. (1990): Optimal release policies for a flexible SRGM, to be published in *Reliability Eng. and Sys. Safety.*

21. Kareer, N., Kapur, P. K. and Grover, P. S. (1990): An *S*-shaped SRGM with two types of errors, *Microelectron Reliab.*, **30**, 1085–1090.

22. Khohgoftaar, T. M. and Woosock, T. G. (1991): Software reliability model selection: a case study, *Proc. Int. Symp. on Software Reliability Engineering*, May 18–19.

23. Mishra, P. N. (1983): Software reliability analysis, *I.B.M. Sys. Jour.*, **22**(3), 262–270.

24. Musa, J. D., Iannino, A. and Okumoto, K. (1984): *Software Reliability*, McGraw-Hill Book Co., New York.

25. Musa, J. D. (Sept. 1975): A theory of software reliability and its application, *IEEE Trans. Software Engg.*, 312–327.

26. Musa, J. D. and Okumoto, K. (1989): A logarithmic Poisson execution time model for software reliability measurement, *Proc. 7th Int. Conference on Software Engg.*, Orlando, Florida, March 26–29, 230–238.

27. Ohba, M. (1984): Software reliability analysis models, *I.B.M. Journ. Rev. Dev.*, **28**(4), 428–443.

28. Ross, S. M. (1985): Software reliability: The stopping rule problem, *IEEE Trans. Software Engg.*, **SE-11**, 1472–1476.

29. Vallee, F. M. and Ragot, A. (1991): Reliability evaluation using NHPP models, *Proc. Int. Symp. on Software Reliability Engineering*, May 18–19, Austin, Texas 157–167.

30. Xie, M. and Zhao, M. (1991): On reliability growth models with simple graphical interpretations, Research Report, Dept. of Mech. Eng., Linkoping University, Sweden.

31. Xie, M. (1989): A generalisation of *JM*-model, in *Reliability 89*, paper 5 Ba 3, England.

32. Yamada S., and Osaki, S. (1988): Some stochastic models for software reliability analysis, *Fourth Int. Symp. Applied Stochastic Models and Data Analysis*, France, 67–78.

33. Yamada, S. and Osaki, S. (1987): Optimal software release policies with simultaneous cost and reliability requirement, *Euro. Journ. Oper. Res.*, **31**, 46–51.

34. Yamada, S., Ohba, M. and Osaki, S. (1983): *S*-shaped reliability growth modelling for software error detection, *IEEE Trans. Reliab.*, **R-32**, 475–478.

35. Yamada, S. and Osaki, S. (1985): Discrete software reliability growth models, *App. Stochastic Models and Data Analysis*, **1**, 65–77.

36. Yamada, S., Ohba, M. and Osaki, S. (1984): *S*-shaped software reliability growth models and their applications, *IEEE Trans. Reliab.*, **R-33**, 289–292.

37. Yamada, S., Ohtera, H. and Narihisa, H. (1986): Software reliability growth models with testing effort, *IEEE Trans. Reliab.*, **R-35**(1), 19–23.

38. Yamada, S., Osaki, S. and Narihisa, H. (1985); A software reliability growth model with two types of errors, *R.A.I.R.O.*, **19**(1), 87–104.

39. Yun, W. Y. and Bai, D. S. (1990): Optimal software release policy with random life cycle, *IEEE Trans. Releaf.*, **39**(2), 167–170.